W9-ACS-847

THE SPIRIT OF SYSTEM

Lamarck in 1802. From *Archives du Muséum d'Histoire Naturelle,* sixth series, *6* (1930) .

THE SPIRIT OF SYSTEM

LAMARCK AND EVOLUTIONARY BIOLOGY

Richard W. Burkhardt, Jr.

HARVARD UNIVERSITY PRESS
Cambridge, Massachusetts, and London, England · 1977

Printed in the United States of America

Library of Congress Cataloging in Publication Data

Burkhardt, Richard Wellington, 1944–
The spirit of system.

Bibliography: p.
Includes index.
1. Lamarck, Jean Baptiste Pierre Antoine de Monet de, 1744–1829. 2. Evolution—History. 3. Biologists—France—Biography. I. Title. [DNLM: 1. Evolution—History. 2. Biology—History. WZ100 L2153BU]
QH31.L2B87 575.01'66'0924 76–53804
ISBN 0–674–83317–1

For my father and mother

Acknowledgments

In writing this book I received the kind help of numerous institutions and individuals. My research was supported by a National Defense Education Act Title IV fellowship; a National Science Foundation travel grant; a Josiah Macy, Jr., Foundation fellowship; and a University of Illinois faculty summer fellowship. I was also aided by the gracious assistance of members of the staffs of various libraries, including the Bibliothèque du Muséum d'Histoire Naturelle, Paris; the Bibliothèque Nationale, Paris; the library of the Institut de France; the library of the British Museum; the University Library, Cambridge; Bodleian Library, Oxford; Widener Library and the libraries of the Museum of Comparative Zoology and the Gray Herbarium, Harvard; and the libraries of the University of Illinois at Urbana-Champaign.

Permission to publish the Lamarck and Cuvier manuscript materials quoted in this book was kindly granted by the Bibliothèque du Muséum d'Histoire Naturelle in Paris, which also permitted the reproduction of the Schutzenberger photoengraving of Thévenin's portrait of Lamarck. The Syndics of Cambridge University Library permitted the publication of the manuscript comment by Darwin which appears in Chapter Six. D. Reidel Publishing Company allowed the republication of materials that I published earlier in the *Journal of the History of Biology*.

The following persons read my manuscript in its entirety at one stage or another of its development: John C. Greene, Roger Hahn,

Ernst Mayr, and Everett Mendelsohn. I am grateful to them for their advice. Thanks is also due to the following friends and colleagues who in different ways contributed to my work: Barbara Buck, Peter Buck, Leslie Burlingame, Paul Farber, Stephen Jay Gould, M. D. Grmek, M. J. S. Hodge, Yves Laissus, Evan Melhado, Barbara Gutmann Rosenkrantz, and Jean Théodoridès. Irene Blenker, Janice Draper Cunningham, and Nancy Edwards helped with the typing of the manuscript. William Bennett, Dennis Davis, and Harry Foster provided invaluable editorial assistance. Ann Blum and Richard Johnson at the Museum of Comparative Zoology, Harvard University, helped me select the illustrations.

My greatest debt is to Jayne Burkhardt. Her contributions have ranged from the very specific (such as ruthlessly pointing out the inadequacies of sentences it took me days to write) to the very general (most notably, tolerating this enterprise for its duration). She takes a skeptical view of authors' acknowledgments and says I should write: "Had it not been for my family, I would have finished sooner." Whatever truth there may be in this statement, in a fundamental way it lacks context. For the special context they have provided me, I am especially grateful to Jayne, Rick, and Fritz.

Contents

FIGURES

THE SPIRIT OF SYSTEM

Introduction

The French biologist Jean-Baptiste Lamarck is not one of the forgotten figures in the history of science. He is typically associated with the idea of the inheritance of acquired characters, exemplified by a Kiplingesque account of how the giraffe got its long neck, and with two notorious episodes in twentieth-century biology: the Kammerer affair, which involved the forgery of scientific evidence, and the Lysenko affair, which involved the repression of scientific thought. These associations have enhanced interest in Lamarck as a historical figure. By and large, however, the ways in which Lamarck has been remembered fail to illuminate either the nature of his own work or the scientific problems of his day. If Lamarck's thought is to be appreciated in its own historical context, certain common assumptions regarding what was central to his thought must be revised.

Lamarck did believe in the idea for which he is most famous: the idea that somatic modifications resulting from an organism's development of particular habits may be passed on to that organism's offspring under the appropriate conditions. This is the idea that has come to be known as the "inheritance of acquired characters." But this idea was neither one which Lamarck originated nor one for which he claimed special credit. What is more, in his own day, he was not criticized for holding it. His contemporaries took exception to his claim that organic change could proceed beyond the limits of the species type, but they did not doubt that within these limits the results of habit tended to become hereditary. For them, as for Lamarck himself, the reality of the inheritance of acquired characters was not an issue.

The reality of the inheritance of acquired characters did not become an issue until three quarters of a century later, in the 1880s, when the German embryologist August Weismann distinguished sharply between the body and germ cells of the organism and questioned whether characters acquired by the body cells could ever be transmitted to the germ cells and thus passed on to the next generation. It was then that the inheritance of acquired characters came to be commonly identified as the "Lamarckian" or "neo-Lamarckian" mechanism of organic change, in contrast to the "Darwinian" or "neo-Darwinian" view that evolution proceeds through the natural selection of small, fortuitous variations. Charles Darwin himself, as his contemporaries were well aware, believed that the effects of habit could become hereditary, and he had granted such inheritance an important role in organic change. When the names were assigned to the theoretical positions, however, this detail was considered negligible. Darwin's position was "purified" and Lamarck's name became indissolubly bound to an idea which was much less a concern of his own than a concern of the biologists of the late nineteenth and early twentieth centuries.

The example of the giraffe, like the idea of the inheritance of acquired characters, looms larger in modern renderings of Lamarck's theory than it did in Lamarck's own. Lamarck used the example to illustrate the effect of habit upon organic form. He maintained that the giraffe's extraordinarily long neck and forelegs were the result of the constant efforts of the giraffe's ancestors to reach upward to browse upon the leaves of trees. But this example was simply one of many he offered to help explain part of the evolutionary process. Indeed, unlike the tortoises and finches of the Galapagos and their apparent importance in stimulating Darwin's evolutionary views, the giraffe seems not to have been a stimulus to Lamarck's evolutionary theory but instead only an afterthought. Lamarck first mentioned the giraffe in 1802 in the index to his *Recherches sur l'organisation des corps vivans* after he had already set out in the main text other examples of the importance of habit for organic change.[1] Lamarck's contemporaries paid less attention to the giraffe example than to the others, recognizing all the while that Lamarck's theory was concerned not only with adaptive change at the species level but also with the origin of life, the different levels of organic complexity, and the faculties associated with specific organic structures. Today, however, the major features of Lamarck's biological theorizing have been forgotten while the giraffe example remains to caricature Lamarck's thinking.

As for the association of Lamarck's name with the Kammerer and Lysenko affairs, this is simply a function of the historical accident through which the inheritance of acquired characters came to be identified as the "Lamarckian" principle. The first of these episodes took place in the first quarter of the twentieth century and centered about the attempt to prove the inheritance of acquired characters experimentally. It came to a dramatic head in 1926, with the announcement that evidence the Austrian biologist Paul Kammerer had offered in support of the inheritance of acquired characters had been faked. Six weeks after this announcement Kammerer shot himself. The second episode extended from the 1930s to the 1960s and involved the virtual destruction of genetics in the Soviet Union under the scientific dictatorship of Trofim Denisovich Lysenko. Profoundly ignorant of the content and methods of contemporary biology, Lysenko espoused one recognizable scientific idea: the inheritance of acquired characters.

Lamarck himself, it should go without saying, is not to be blamed for either the Kammerer or the Lysenko affair. Nor was Lamarck's situation comparable to that of either Kammerer or Lysenko. The guiding issues of Lamarck's work were not the guiding issues of the work of the two most famous "Lamarckians" of the twentieth century, and though each of the three cases involved a kind of "politics," Lamarck's position in the scientific community differed significantly from Kammerer's or Lysenko's.

The debate over the inheritance of acquired characters in which Kammerer was involved would have taken place whether or not Lamarck had ever written upon organic evolution. This debate arose in the last decades of the nineteenth century as a natural response to the problem of the source of heritable variations, a problem left unresolved by Darwin in his *Origin of Species*. At the beginning of the twentieth century the debate was still going on. Neither Weismann's theoretical arguments nor his often cited experiment involving the amputation of mouse tails for several successive generations had sufficed to overcome the widely held belief that at least some kinds of environmentally induced modifications could be transmitted from one generation to the next. The neo-Lamarckians were able to call upon a wide variety of circumstantial evidence in support of their position. Until the twentieth century, however, they could cite little experimental evidence in their favor. Not that they felt entirely compelled to do so, since naturalists of the nineteenth century had, by and large, not regarded experimentation as a key to resolving the major issues of organic evolution. But as experimentation came to be increasingly lauded for its importance to biology, the debate over the inheritance of acquired characters was inevitably dragged into the experimental arena. When, in the first decade of the twentieth century, Paul Kammerer came forward with claims of having proved the reality of the Lamarckian principle experimentally, his claims were naturally greeted with enthusiasm by the neo-Lamarckians. It was a staggering blow to the neo-Lamarckian cause in 1926 when the nuptial pads that were purported to be the inheritance of an acquired character in Kammerer's only remaining specimen of "midwife" toad were found instead to be injections of India ink under the toad's skin. Kammerer's suicide shortly thereafter was widely interpreted, correctly or not, as an admission of guilt.[2]

By the time Lysenko began endorsing the idea of the inheritance of acquired characters, the idea had been widely, if not universally, discredited. This general disenchantment with the Lamarckian principle was the result of a combination of factors: the lack of clear experimental evidence that acquired characters are ever inherited, the lack of a plausible mechanism by which modifications of somatic cells might be transcribed in the germ cells, the taint of fraud resulting from the Kammerer case, and the availability of alternative explanatory frameworks that seemed to render reliance on the Lamarckian principle unnecessary. Nonetheless, Lysenko advanced the battle-worn idea that environmentally induced modifications could be inherited, seeing in it the promise that man could direct organic change. He explained his dissatisfaction with genetics as it had developed in the West in the following terms: "Mendelism-Morganism is built entirely on chance; this 'science' therefore denies the existence of necessary relationships in living nature and condemns practical workers to fruitless waiting. There is no effectiveness in such science. With such a science it is impossible to plan, to work toward a definite goal; it rules out scientific prediction."[3]

Lysenko's rise to power has been shown to have been less a function of his biological theories or of Marxist philosophy than of the needs of Soviet agriculture.[4] Lamarck, interestingly enough, believed in one of the particular phenomena that brought attention to Lysenko a century and a half later: the alleged conversion of winter wheat into spring wheat.[5] Lamarck, however, never appealed to the practical implications of the inheritance of acquired characters—either agricultural or social—as a reason for believing in such inheritance.

Many social theorists have seen in the inheritance of acquired characters a possible foundation for social progress. In the late eighteenth century, Condorcet assumed the inheritance of acquired characters in constructing his vision of the social and organic improvement of mankind. In the nineteenth century Herbert Spencer, Friedrich Engels, and Lester Ward based their respective views of human progress on the Lamarckian principle. In the twentieth century Kammerer did not hesitate to point to the advantages that would accrue to the human race if acquired characters were inherited and if the proper reforms were instituted in society. But Lamarck never expressed such concerns. This may have been because he felt organic change operated too slowly to be effective on a human time scale. It may also have been because he was not particularly interested in social change or because the political events of his day left him skeptical about the prospects of far-reaching social change being achieved. Whatever the case, there is no reason to consider Lamarck an early Lysenko, seeking to dominate both men and nature on the basis of a particular view of nature's operations.

But if Lamarck is not to be seen as an early Kammerer or Lysenko—or for that matter as an early Darwin (another perspective which has served to obscure fundamental aspects of Lamarck's thought)—how is he to be seen? If his story does not revolve about the idea of the inheritance of acquired characters, be it in the form of giraffes' necks, toads' nuptial pads, or the tails of mice, if he was not resisting the objectification of nature or espousing a social philosophy, and if he was not attempting to write an *Origin of Species*, what were his basic concerns?

Lamarck was active as a scientist in the late eighteenth and early nineteenth centuries, a period in the history of science when the pursuit of knowledge was becoming increasingly specialized but individuals did not necessarily feel constrained to limit themselves to a single field of study. Lamarck distinguished himself first as a botanist and then as an invertebrate zoologist, but he also took it upon himself to ponder fundamental questions regarding physics, chemistry, meteorology, geology, and a new science he called "biology." His attempts to lay conceptual foundations for these various fields were received far less enthusiastically by his contemporaries than were his efforts as a classifier of plants and animals. Nonetheless, he regarded his general observations on nature's operations and science's needs to be more valuable than his work as a systematist. He was especially proud of his thinking with respect to his new science of "biology," which though encompassing "everything relating to living things" was especially concerned with the organization of living things and the growing complexity of this organization as the result of the prolonged action of "life's motions."[6] He felt that in studying "the origin of living bodies and ... the principal causes of the diversity of these bodies and the developments of their organization and faculties" he had tackled not only "the most vast and important goal that one can embrace in the study of nature" but also the problem "that is undeniably the most difficult to resolve."[7] Privately, he suggested that his own work in biology would be comparable to Newton's in astronomy.[8] The guiding idea of his biology was that nature had first brought into being the very simplest forms of life and then from these, with the aid of favorable circumstances and a great deal of time, successively produced all the different forms of life on earth.

Diverse opinions have been expressed concerning the nature and merits of Lamarck's work. Early in the nineteenth century Lamarck was appreciated more for his systematic than for his speculative endeavors. Later in the century, once the idea of organic evolution gained general acceptance among biologists, he was typically portrayed as a great thinker whose misfortune was to have lived at a time that was not "ripe" for the appreciation of his ideas. Biographers of Lamarck at the turn of the century characteristically set themselves the dual task of rescuing Lamarck from the obscurity into which he had been cast by Cuvier and promoting the cause of the neo-Lamarckian explanation of evolution.[9] In contrast, the American historian of science Charles Gillispie has seen Lamarck not as the insufficiently appreciated founder of modern evolutionary theory but rather as a thinker whose "theory of evolution belongs to the contracting and self-defeating history of subjective science."[10] Gillispie has written that Lamarck failed to accept the trend of modern science—to objectify nature by mathematizing it—and that Lamarck, like Lysenko, sought a morality based on nature and thereby invited the domination of science by politics. Gillispie's service in indicating that Lamarck might best be understood in some other role than precursor to Charles Darwin has been invaluable. But the particular perspective in which Gillispie put Lamarck's work has been less fortunate. Presenting Lamarck as an object lesson in the difference between good and bad science is little better than treating him as a thinker too far ahead of his time. Neither approach has shed much light on the scientific issues of Lamarck's time or the maneuverings through which different individuals, ideas, and styles of scientific thought were promoted or held in French science at the beginning of the nineteenth century.

Since Gillispie's work in the 1950s, Lamarck has received increasing attention from historians of science.[11] Lamarck's writings have been reread with some care, and new attempts have been made to evaluate his biological thought on its own terms and in its proper historical context. But there has not been a systematic and detailed study of Lamarck's biological thought in general or evolutionary thought in particular which addresses itself to the range of questions that need to be confronted if Lamarck's place in the history of biology is to be assessed satisfactorily. What role did Lamarck, as a self-styled "naturalist-philosopher," see himself playing? How was his "biology" related to the natural history of his day and to the other fields of science in which he took particular interest? What were the conceptual foundations of his early scientific thought, and in what ways did the evolutionary views he eventually advanced represent a development upon or rejection of these foundations? What was the relation of his evolutionary thought to various eighteenth-century ideas about the origins of organic diversity? What were his personal and intellectual relations with his colleagues, and what was his status within the scientific community during the course of his career? What brought him to a belief in organic mutability in the first place? What were the basic features of the theory of evolution he eventually developed? And how was his theory of evolution received in its own day?

Equal weight will not be given here to each of Lamarck's diverse intellectual pursuits. Lamarck's thoughts on physics, chemistry, and meteorology will be examined primarily insofar as they shed light on the general patterns of Lamarck's scientific thought and particular features of his natural history and biology. Lamarck's physico-chemical and meteorological systems will be considered, but the treatment of them will be relatively brief. It is the author's judgment that Lamarck's fascination as an historical figure lies in his work as a naturalist and biologist and that since Lamarck himself viewed his biological studies as constituting the most interesting part of his work the focus of the present study is justified.

Lamarck poses special difficulties for the biographer. No autobiography, diary, or correspondence exists to provide a close view of Lamarck's personality or intellectual development. His published and manuscript works are all devoted to scientific subjects. If passions manifest themselves in his writings, they are primarily the passions of the naturalist-philosopher and academician addressing his attentions to nature on the one hand and his colleagues on the other. From these writings one learns a little about his habits (he arose at five o'clock every morning to take the first of three daily readings of the temperature, atmospheric pressure, and wind direction) and about his health (he found his energies insufficient to support an active physical life and an active mental life at the same time, especially after 1800).[12] But his writings disclose very little with respect to his family life, his personal relations with particular colleagues, or his activities during the French Revolution.

Some bare biographical data are available.[13] Lamarck was born at Bazentin in Picardy on August 1, 1744, the last of eleven children in a family of noble lineage but modest resources. As the youngest in the family, he found himself destined not for the military, which had provided the careers of his brothers, father, and several generations of forebears, but rather for the priesthood. He attended for a time the Jesuit College at Amiens, but following his father's death in 1759 and the expulsion of the Jesuits from France in 1761 he was no longer restrained from joining the army, and he proceeded to enlist. He pursued a military career from 1761 until 1768, when an injury forced him to resign. Upon leaving the army he spent some time at his family's home at Bazentin and then went to Paris, where he eventually embarked upon a scientific career.

The basic outline of Lamarck's career is known and will be set out in later chapters. We also know that Lamarck had six children by one woman before he married her on her deathbed in 1792; that he later had at least two more wives and two more children; that one of his sons died of yellow fever, a second was deaf, and a third was insane; that he himself was completely blind, as well as impoverished, for the last eleven years of his life; that his daughter Rosalie stayed with him and cared for him until his death in 1829; and that after his death, his son Auguste criticized him for failing to attend to the needs of his family.[14]

A complete and vivid portrait of Lamarck as a person cannot be drawn from the writings of Lamarck's contemporaries any more than it can be drawn from Lamarck's scientific studies. His contemporaries' writings contain only fleeting glimpses of him, and the brief obituary notices that followed his death tend to reveal more about their authors than they do about Lamarck. The few incidents in which Lamarck's behavior has been recorded show Lamarck to have been like most mortals: capable of pettiness as well as generosity, of frailty as well as strength. One visitor to Paris was taken aback by the way Lamarck "always took occasion to attack, with violence, what he knew to be my most favourite sentiments."[15] But other naturalists discovered Lamarck to be entirely gracious and quite willing to go out of his way to help them when they needed to examine his collections.[16] Augustin-Pyramus de Candolle, seeking election to the prestigious first class of the Institut de France (the successor to the Académie des Sciences) was dismayed to find that when it came to filling a vacancy in the botany section Lamarck succumbed to pressure from Fourcroy and failed to vote for de Candolle, even though Lamarck regarded de Candolle's work to be superior to that of the rival candidate.[17] Yet Lamarck seems to have been generally stalwart in maintaining his own opinions, even when they met with opposition, contrary evidence, ridicule, or neglect.

In regard to Lamarck's tenacity in maintaining a position irrespective of its popularity (or for that matter, its wisdom), a familiar anecdote bears retelling. The story concerns the role Lamarck played in his first military engagement: the battle of Fillinckhausen of July 14, 1761. Just how true the story may be scarcely matters. The story was related to Georges Cuvier by Lamarck's son, Auguste, following his father's death, and since Auguste de Lamarck's account presumably approximates the story Lamarck told to his children, it evidently reveals what kind of person Lamarck considered himself to be. Lamarck's first battle proved to be a losing cause. Upon arriving at the front, he was placed with a company of grenadiers that came under heavy fire and soon lost all of its commanding officers. As the rest of the French army retreated, Lamarck and his comrades were overlooked and consequently received no instructions to withdraw. When the worried grenadiers turned to the young Lamarck for advice, he reportedly told them: "We have been placed here, we must not leave our post without having been relieved. If you fear being captured, leave. As for me, I will stay." The company stayed and was finally relieved later in the day.[18] This story has been recounted frequently by Lamarck's biographers as an indication of his bravery and loyalty as a young man. An analogy between Lamarck's military and scientific behavior seems obvious: Lamarck was courageous in the face of strong opposition. In fact, however, an analogy of a much tighter sort can be drawn. It will be argued in the following pages that it was in no small measure Lamarck's refusal to give up certain scientific and philosophical positions—positions that in the late 1790s were under fire and of questionable defensibility—that led him to his belief in evolution in the first place.

One's sympathies inevitably go out to the aged Lamarck: blind, poor, and unappreciated by his contemporaries. This lonely figure stands in stark contrast to his major critic, Cuvier, whose talents, both scientific and political, earned for him a position of great prestige and power in French science in the first three decades of the nineteenth century. Yet if Lamarck appears as a sympathetic figure, especially when compared to Cuvier, Lamarck's personality does not seem to have been one which would have readily endeared him to others. Lamarck appears to have been reserved, prone to a certain bitterness and self-righteousness, and not especially receptive to the ideas of others though very desirous of being taken seriously himself.

Lamarck was a sensitive person, susceptible of being wounded deeply by personal affronts. When traveling through Europe with Buffon's son in his charge, the spoiled young Buffon grew to hate being under Lamarck's surveillance and poured ink all over Lamarck's linens and best clothing. In the last years of Lamarck's life it still grieved him to recall the incident.[19] In 1802, while reading a memoir on meteorology to the first class of the institute, Lamarck was interrupted by scathing criticism from Laplace.[20] Perhaps for this reason Lamarck never read a memoir to the institute again. More painful still, undoubtedly, was the rebuff Lamarck received from Napoleon when he sought to present his *Philosophie zoologique* to the Emperor. Napoleon rudely dismissed Lamarck, mistaking the new book for one of Lamarck's meteorological annuals, which Napoleon considered unworthy of a member of his Académie des Sciences. The elderly zoologist was reduced to tears.[21]

Though it is unfortunate that so little is known of Lamarck's personal life, it does seem that his scientific work was the focus of his existence. He found an escape from his personal misfortunes in the solitude of his studies. As he reflected on the broad panorama of nature's operations, the anguishes of his life were soothed. The neglect of his views, the painful personal affronts, the physical infirmities that weighed increasingly upon him, and the financial and emotional problems posed by his large family all shrank in importance as he contemplated the laws of the cosmos.

Much more can be said about Lamarck's scientific ideas than about his personal life. Nonetheless, in the intellectual realm as well as in the personal, too little is known about Lamarck. He rarely stopped to review the ways his thoughts had developed on any subject. When he did so, he overrationalized the progress of his thinking. And more often than not he failed to identify the individuals from whom he borrowed ideas.

Two contrary tendencies have emerged in historical treatments of Lamarck's views as the result of Lamarck's inattention to the sources of his views. One has been to presume that Lamarck was extraordinarily independent in his thinking. The other has been to propose a whole host of thinkers as his precursors on the basis of a greater or lesser resemblance of his views to theirs. There is no question that certain of Lamarck's views were less novel than has frequently been supposed. An individual's thinking is not explained, however, simply by identifying contemporary or earlier thinkers who held the same or similar views. Too great a concern with Lamarck's precursors (and his successors as well) has led to a failure to analyze his own intellectual development and the dynamics of the scientific community of which he was a part. Obviously, however, the quality and originality of Lamarck's thought cannot be judged solely by studying that thought alone. Comparing his thought with that of his immediate predecessors and close colleagues is essential in identifying the ideas (and combinations of ideas) that were considered viable in his time.

Reconstructing the intellectual milieu in which Lamarck operated as a means of evaluating the guiding features of Lamarck's evolutionary thought is one of the major tasks of this book. Instead of placing Lamarck on a continuum between Buffon and Charles Darwin, or worrying about what, if anything, Lamarck owed to Charles Darwin's grandfather, Erasmus, Lamarck's work will be studied here by comparing it above all to the works of his colleagues.[22] Lamarck quite evidently owed a good deal to certain of his predecessors. The ideas of Buffon, Linnaeus, and Von Haller, for example, reappear with considerable frequency in Lamarck's writing. Yet the dynamics of Lamarck's thought are perhaps best appreciated by comparing his thinking not so much to those who came before him as to those of his time. In this regard his views must be compared not only with those of such well-known contemporaries as Daubenton and Cuvier but also those of such lesser lights as Bosc, Bruguière, Duchesne, Faujas de Saint-Fond, Lacépède, Olivier, and Reynier.

The idea of organic evolution may be viewed as a natural outgrowth of certain trends in eighteenth-century thought, such as an increasing materialism or the development of historical thinking. But this leaves one with little notion of why Lamarck came to develop his theory of evolution at the time that he did or why his contemporaries were by and large unsympathetic to his views. If Lamarck's place in the history of biology is to be understood, the common assumptions of the naturalists of Lamarck's day and the theoretical and methodological issues that divided these thinkers must be known. What counted most for Lamarck himself must also be identified. An individual with considerable pride, Lamarck sought to define for himself a special scientific role—that of naturalist-philosopher—at a time when his contemporaries were inclined to be skeptical both of the utility of such a role and Lamarck's own ability to fill it. He was not merely a precursor of Darwin or the culminator of various trends in eighteenth-century thought. Seeking to understand the essence of life and its various manifestations, responding not only to general developments in the sciences but also to the specific demands of his own work as a naturalist, Lamarck reveals himself as very much a man of his time.

CHAPTER ONE

Jean-Baptiste Lamarck, Naturalist-Philosopher

In the first years of the nineteenth century, as he neared his sixtieth birthday, Jean-Baptiste Lamarck felt himself to be at the peak of his intellectual productivity. A distinguished career in botany lay behind him. A new career in invertebrate zoology was bearing further testimony to his talents as a systematist. And the project he saw as his crowning glory was taking shape in his mind: he intended to set out for his contemporaries an entire "terrestrial physics," encompassing "considerations of the first order" relative to the fluctuations of the earth's atmosphere (meteorology), the changes of the earth's surface (hydrogeology), and the origin and nature of living things (biology).[1] He was, however, experiencing a difficulty. His contemporaries were proving unreceptive to his new ideas.

In 1797 Lamarck had told his colleagues in the first class of the Institut de France that the earth's atmosphere exhibited regular fluctuations corresponding to the moon's elevation above or declination below the earth's equator.[2] Two years later he had told this same group that the present features of the earth's surface were the result of everyday kinds of processes that had operated over long periods of time.[3] In 1800 he had told his students at the Muséum d'Histoire Naturelle that the phenomena of life could be explained in a wholly naturalistic fashion: the origin of living things, their diversity, and the special faculties they displayed were no longer mysterious once nature's true way of proceeding was recognized.[4]

To Lamarck's dismay at the beginning of the new century, these and other of his major ideas were being ignored. Moreover, his contemporaries were conveying to him the impression that not only did they consider certain ideas of his to be wrong, they regarded his basic approach to science to be retrogressive. To Lamarck, who prided himself on the special intellectual role he felt he was playing, this last judgment was the unkindest of all. As he reflected on the dominant mood of the scientific community his feelings became anguished and bitter. While attending the meetings of the leading scientific societies, he imagined that winks and smiles were being exchanged behind his back by conspirators who had combined to repress his novel views.

Lamarck had not always felt so estranged from the rest of the French community. In earlier years French science had had its interest groups, centers of power, and intrigues. In the 1770s and 1780s, however, Lamarck was a man on the way up. Others perceived him as representing a modern trend in contemporary botany and they supported him. During the most difficult years of the French Revolution he was an active figure in French scientific circles, and he emerged from that troubled period with a well-established position in the official scientific structure. In the mid-1790s, however, the events which fed his paranoia of the turn of the century began unfolding. In order to appreciate the successes and failures of Lamarck's career, it is necessary to understand the enterprise of natural history in his day, the conceptual and methodological concerns of his contemporaries, the institutional frameworks within which he operated, and the kind of thinker he perceived himself to be.

IN FRANCE in the second half of the eighteenth century there was an immense popular enthusiasm for natural history and at the same time an increasing concern on the part of serious naturalists regarding their methods, goals, and the precise boundaries of their field.

Just how fashionable natural history was in France in the period in question is suggested by a variety of indices. Buffon's *Histoire naturelle* and Pluche's *Spectacle de la nature* graced the shelves of private libraries even more frequently than did Voltaire's *La Henriade* or Rousseau's *La Nouvelle Héloïse*.[5] When the first three volumes of the *Histoire naturelle* appeared in 1749, the edition, though a relatively large one, sold out in six weeks.[6] By mid-century the height of fashion in French society was no longer the collection of antiquities but rather the possession of such apparatus as electrostatic generators and Leyden jars (for demonstrating the wonders of the new experimental physics) and natural history *cabinets* (for displaying the diversity, beauty, and curiosity found in nature's three kingdoms).[7] Natural history cabinets became the prized possessions not only of naturalists, physicians, and apothecaries but also of businessmen, military officers, lawyers, ecclesiastics, ministers of state, and members of the royalty.[8] Merchants did a lucrative business supplying collectors with specimens for their cabinets, so much so that at the end of the century some six hundred merchants were making a living in Paris selling shells to their clientele.[9] Courses in natural history offered at the Jardin du Roi and privately by Valmont de Bomare were well attended.[10] "Herborizations"—brief excursions into the countryside to study plants in their natural habitats—also became popular.[11] Promoted in no small measure by Jean-Jacques Rousseau's *Lettres sur la botanique,* herborizations provided the body with exercise and the soul with a chance to commune with nature.[12] Numerous expeditions were undertaken by naturalist-voyagers to unexplored parts of the world. In contrast to herborizations, these expeditions were genuinely perilous, so much so that one naturalist at the end of the century suggested drawing up a "martyrology of savants" and maintained that naturalist-voyagers would play the dominant role in it. That individual naturalists undertook the hazardous adventures they did indicates the appeal of natural history in this period.[13] A comparable indicator is the fact that in 1794, when the armies of France passed beyond their country's borders, the National Convention sent two professors from the Muséum d'Histoire Naturelle with them. As members of a "Commission on the Sciences and the Arts," the function of these professors was to requisition objects of science and the arts for "the French people." Prominent among the objects they appropriated and shipped back to France were natural history specimens.[14]

In the 1750s, L.-J.-M. Daubenton, then Buffon's major collaborator on the *Histoire naturelle,* observed: "In the present century the science of natural history is more cultivated than it ever has been; not only do the majority of men of letters make it an object of study or relaxation, but there is in addition a greater taste for this science spread throughout the public, and each day it becomes stronger and more general."[15] Daubenton wondered whether natural

A naturalist's museum of the seventeenth century, illustrating various curiosities of natural history. The possession of natural history collections became extremely fashionable in France in the eighteenth century. From O. Worm, *Museum Wormianum,* Leiden, 1655.

history would in time lose its popularity to another science or another activity. As it turned out, without precluding interests in other things scientific—among which were such marvels as electricity, "animal magnetism," and balloon ascents—the general enthusiasm for natural history remained high to the end of the century.

The motivations behind the late-eighteenth-century interest in natural history were as diverse as the expressions of this interest. Beyond the promptings of fashion and the passions of the collector, one may readily identify religious, moral, psychological, philosophical, and utilitarian motives that contributed to the study of nature's productions. Often, more than one of these inspirations found expression in the writings of a single individual. Linnaeus, for example, at various times discussed or exemplified all of them. What seems especially prominent in Linnaeus' writings, as compared with the writings of the major French naturalists of the late eighteenth century, is the perception of classification as a respectful ordering of God's Creation. Combining piety and vanity, he confidently observed of himself: "God has permitted him to see more of his created work than any mortal before him. God has bestowed on him the greatest insight into nature study."[16] Several French naturalists achieved equal heights of immodesty. By and large, though, they fell short of Linnaeus in piety. Réaumur was notable for the way his studies of nature inspired him to praise the Creator, but this kind of religious enthusiasm is lacking in the writings of Buffon, Daubenton, Adanson, Lamarck, and Cuvier.

Linnaeus served as more than just an example of the religious incentives to study nature. He was also proof that discoveries in natural history could make a person famous. But if some naturalists hoped to make themselves known to the world through their studies, others, like Rousseau, saw natural history primarily as a refuge from the world's cares.[17] Still others, like Bernardin de Saint-Pierre, saw natural history as a source for moral lessons through which individuals and society might be made better.[18] Lamarck came to regard the most important aspect of natural history as that part which appealed to the investigator he called the *naturaliste philosophe,* or naturalist-philosopher. Pursuing knowledge basically for its own sake, he sought to understand the natural affinities among nature's productions and the general processes by which nature operated. He viewed this as the noblest activity the student of nature could define for himself. Daubenton, in contrast, was more skeptical of man's

Conchology came to the forefront of the enthusiasms of the collector-naturalist in the eighteenth century. Pictured here is the frontispiece of the third edition of Dezallier d'Argenville's *Conchyliologie* (1780).

ability to perceive the "natural order" of things, and he treated natural history more as a source of potentially useful discoveries than as an end in itself. This utilitarian motive must not be underrated. Plants, in addition to their fundamental importance as a source of food and fibers, had long been valued for their specific medical uses. Indeed, on into the eighteenth century botany was perceived primarily in the role of handmaiden to medicine. Animals likewise had their obvious utility, and naturalist-voyagers typically justified their expeditions with the prospect of discovering new plants and animals of economic value or at least new sources of plants and animals of economic value. The Dutch monopoly of the lucrative spice trade was an instructive example to other countries. When, in the 1770s and 1780s, the French managed to import and cultivate spice trees in a few colonies of their own, they were naturally pleased with themselves and began envisioning greater things to come.[19] During the French Revolution, when the status of the various French scientific institutions was placed in jeopardy, the naturalists at the Jardin du Roi were able to claim that their institution was of special value, and the popular acceptance of this claim undoubtedly contributed to the Jardin's preservation and reconstitution while other institutions were abolished.

THIS MUCH SAID about the breadth of interests and motives involved in natural history in the late eighteenth century, it is worth looking more closely at how a professional defined his field. For this purpose, an excellent example is the naturalist and comparative anatomist L.-J.-M. Daubenton. Daubenton was keeper and demonstrator of the cabinet of natural history at the Jardin du Roi, collaborator with Buffon on the early volumes of the *Histoire naturelle,* occupant of the first chair of natural history in France (created for him at the Collège de France in 1778), holder of additional professorships at the veterinary school of Alfort, the École Normale, and the Muséum d'Histoire Naturelle, member of the Académie des Sciences and the Institut de France, and so forth.[20] His fame did not equal that of Linnaeus or Buffon, but to some of his fellow naturalists he appeared as an appropriate corrective to the excesses of the two giants of eighteenth-century natural history. Although more wary of speculation and more precise in his anatomical observations than Buffon, he refused to align himself with the Linnaean classifiers whose enterprise appeared to him dry and incomplete.

L.-J.-M. Daubenton (1716–1800), the Nestor of French natural history when the Jardin du Roi was transformed into the Muséum d'Histoire Naturelle during the French Revolution.

Daubenton always took care to distinguish natural history from related fields. A full natural history, he acknowledged, would include the study of all physical things, including the air, meteorological phenomena, and heavenly bodies. These last subjects, however, were commonly understood to be separate from natural history, and so too, he felt, should be all those fields of science "which do not represent their objects in a state of nature."[21] Natural history, in other words, did not include chemistry, metallurgy, agriculture, technology, *materia medica,* anatomy, or medicine. As he explained, when the structure of a mineral or the organization of a plant or animal is altered through a technical process, the specimen in question ceases to be the concern of the naturalist. Daubenton proposed a division of scientific labor. It was the chemist's task to pulverize, dissolve, macerate, distill, and vitrify nature's productions. The metallurgist's role was to extract metals from mines. The agriculturalist helped nature produce plants by tilling and fertilizing the soil. The dyer and the pharmacist extended the properties of substances, making them more active. The anatomist unveiled the tiny, hidden structures of living things. The physician studied the mechanisms of the human body and learned to repair the body when it was out of order. The naturalist was left to contemplate "the minerals, plants, and animals in their different states, without mixing the processes of *art* with the operations of nature."[22]

The naturalist's job, as Daubenton thus defined it, was restricted but still immense. Natural history, in his view, was by no means synonymous with the naming and classifying of species. Indeed, he expressed a wish to distinguish botany, as commonly practiced, from natural history, because contemporary botanists seemed to be primarily concerned with botanical nomenclature and classification and for Daubenton this was not the most important part of the study of plants.[23] He supposed that the first object of the naturalist was to learn how to recognize nature's productions and distinguish them from one another, but he saw this as only the initial step in the naturalist's education. The complete naturalist was an individual who confronted nature's productions in all their aspects: "[He] considers the brute and organized bodies in the different states through which they successively pass from their formation or birth to their destruction. He describes their structure and organization. He tries to discover their origin and the causes of the changes which occur in the course of their existence."[24]

The major issues that confronted the naturalist of the late eighteenth century find expression in Daubenton's various discussions of natural history: the proper handling of description and nomenclature ("the greatest mistake [naturalists] have made is the multiplicity of names for the same thing");[25] the utility of methods of classification and whether such methods represent divisions actually existing in nature ("all these methodical divisions into orders, classes, and genera depend on the whim of the naturalist who imagines them: they are not determined by the nature of things");[26] the principal differences among the productions of nature (the most important being the distinction between "brute" [inorganic] bodies on the one hand and "organized" bodies on the other); the possibility of arranging nature's productions in a single chain of being (which Daubenton denied); the existence of gradations connecting the minerals, plants, and animals or the different classes within these divisions (which Daubenton also denied); the distinction between characteristics which identify species and those which only identify races or varieties; the importance of comparative anatomy in the study of organized beings; the circulation of substances in the overall economy of nature; the value of cabinets of natural history; the contributions of naturalist-voyagers; the dangers of unchecked speculation; the practical applications of natural history; and so on.

The shifts in emphasis that distinguish Daubenton's later accounts of natural history from his earlier ones include a slight lessening of hostility toward methods of classification (despite a continued insistence that such methods were ultimately artificial), an increasing objection to the idea of a single chain of being linking nature's productions (an idea which was finding enthusiastic expression in the writings of the Genevan naturalist Charles Bonnet), and a tendency to drop the idea that nature is divided into three kingdoms in favor of the idea that there are only two basic kinds of natural production: the brute and the organized. In 1755, in the *Encyclopédie* of Diderot and d'Alembert, he drew no distinction between brute and organized bodies but simply stated that animals, plants, and minerals were the basic parts of natural history.[27] In 1782, in the *Encyclopédie méthodique,* he wrote of "the three kingdoms of nature," though he observed: "It is generally agreed that the principal difference between the productions of nature is that some are only brute while others have organs."[28] By 1795 he told his students at the École Normale that the word "kingdom" was altogether out of place in natural history: it was "improper" and "unintelligible."[29] This observation was not made independently of the revolutionary sentiments of his audience. He had already drawn an enthusiastic response from his listeners by observing in an earlier lecture: "The lion is not the king of animals. There is no king in nature."[30] Whatever the immediate political context of these remarks, however, there is a broader conceptual significance to be found in his denial of the existence of three comparable kingdoms of nature. In combining plants and animals in one category of organized beings, set off from the objects of the inorganic world, Daubenton, like many others of his time, was taking a critical step toward the notion of a science of biology.

THE COINER of the word "biology," or at least one of the coiners of that word, was Jean-Baptiste Lamarck.[31] He introduced the word in 1802 to designate that part of his terrestrial physics which was to deal with "considerations of the first order . . . relative to the origin and to the developments of organization of living bodies."[32] When he began writing his "Biologie" he subtitled it "considerations on the nature, the faculties, the developments, and the origin of living bodies."[33] Stated as such, his subject did not represent a distinct break with the concerns that certain naturalists and anatomists had exhibited late in the eighteenth century, though the special views he had developed on the *origin* of living things did set him apart from the vast majority of his predecessors and contemporaries. All this, however, was several decades in the future at the time Lamarck took up the study of natural history.

Lamarck became interested in botany as a young man stationed with the army in the south of France. Equipped with a botanical handbook, the *Traité des plantes usuelles* of Chomel, he herborized in the countryside around Toulon, where his company was garrisoned from 1764 to 1765. After his company was transferred to the Fort of Mont-Dauphin, he had the opportunity to contrast the Mediterranean flora of Toulon with the alpine flora of his new surroundings. An injury put an end to his military career in 1768. He subsequently worked as an accountant at a bank, contemplated the idea of becoming a musician, enrolled in medical school, and then, in the mid-1770s, attempted to make a full-time career for himself in the natural sciences. He became a member of the circle of botanists and students at the Jardin du Roi. It was there that he announced that he could provide a more efficient means of identifying French plants than any currently in use, and furthermore, that he could have his system ready in a year's time. Spurred on, it seems, by at least a verbal bet, he set to work on his "flora of France" and completed it within the designated time.[34]

The possibility of a significant future as a botanist materialized for Lamarck in 1777 when Buffon, impressed by the non-Linnaean aspects of Lamarck's work, arranged to have the work published at government expense with the proceeds from the sales going to the author. Lamarck's cause was also supported at this time by two influential individuals from his home province: le comte de la Billarderie and le comte d'Angiviller. Shortly thereafter Lamarck was chosen (again with Buffon's support) to fill a vacancy in the botanical section of the Académie des Sciences. Under the circumstances, the selection was extraordinary if not entirely unprecedented. The academy had ranked Lamarck as its second choice, behind the physician-botanist Jean Descemet, as the result of an election in which Descemet received twelve votes and Lamarck ten.

By the decision of the King, however, prompted by Buffon and d'Angiviller, the vacant position went to Lamarck.[35]

Several observations must be made about the sequence of events that brought Lamarck in just a few years to a position of note in the French scientific community and to membership in the most prestigious institution of French science in the eighteenth century. The first is that his initial reputation came as the result of his invention of a novel approach to a difficulty that the botanists of his day had not resolved. Was it possible, he asked in his *Flore françoise,* that the same means could be used on the one hand to discover the names of the plants one wished to identify and on the other hand to represent the true affinities between the plants? He answered with a firm negative. The task of finding the name that botanists had given to a plant, he insisted, had to be divorced from the task of determining the place the plant occupied in the "natural order" of the plant kingdom.[36] The first of these tasks was the one on which he concentrated his attentions in the *Flore françoise.* There he set forth a method of "analysis" that was unabashedly "artificial" and that had as its "unique object" the discovery of the names of observed plants. The effectiveness of this method, which involved a system of dichotomous keys, was a major reason for the *Flore*'s success.

The second observation to be made about Lamarck's rise to prominence regards how much he profited from being perceived as someone who represented a modern trend in natural history. Lamarck's attacks on the unnecessary complexity of Linnean systematics in general and the Linnean idea of the reality of genera in particular could only have delighted Buffon, who lacked the botanical knowledge to challenge Linnaeus in the great Swedish naturalist's own major field.[37] It may also have been to Lamarck's advantage that he, unlike his rival Descemet, was not a physician. It was not that the naturalists of his day disapproved of training in medicine. Most botanists then possessed medical degrees. It was coming to be felt, however, that such sciences as botany, chemistry, and anatomy were worthy of study in their own right, not just as adjuncts to medicine. Descemet may have been seen as representing the older, more amateurish approach to botany. At any rate, his membership in the Faculté de Médecine, if not his approach to botany, appears to have earned him the hostility of Tessier and Desfontaines, scientifically oriented botanists belonging to the medical faculty's rival, the Société Royale de Médecine. Some of the votes cast for Lamarck are perhaps best interpreted as votes against Descemet.[38]

It clearly must not be assumed that Lamarck, by being elected to the botany section of the academy, stood out as the new protégé of a happy, united family. Jealous of their specimens and their taxonomic systems, the botanists of late-eighteenth-century France did not work in a spirit of complete mutual respect and cooperation. As one observer of the Paris scientific scene remarked in 1791, "The botanists and amateurs here are anything but communicative; when someone has a rare plant in flower, it is guarded more closely than a treasure."[39] Lamarck's major effort of the 1780s, his contribution to the botanical part of the *Encyclopédie méthodique,* was reportedly spoken of with universal scorn by the scientific botanists of France late in that decade.[40]

Lamarck's most obvious talent as of the late 1770s was not his insight in approaching the more profound questions that the French naturalists of his time were confronting but instead his skill in identifying plants and in describing clearly and concisely the distinguishing characters of species and genera. His thoughts on plant anatomy and physiology were not novel. He had not given the question of "natural method" the attention that Adanson or the Jussieus were giving it. His *Flore françoise* does begin with a long, theoretical discourse, but just how much of this discourse represents his own thinking is uncertain, for under Buffon's instructions, the preparation of the discourse was to be overseen by Daubenton, who proceeded to delegate some of this responsibility to l'abbé Haüy. It was Haüy, Lamarck admitted, who was the source of some of the thoughts on natural method that appeared there.[41]

But if Lamarck had not fully probed the depths of nature study by the time the *Flore françoise* appeared, his concerns were certainly broader than that work revealed. He first became known to the scientific establishment not as a botanist but as a meteorologist, when in 1776 a memoir of his "on the principle phenomena of the atmosphere" was read to the academy.[42] He wanted to expound his thoughts on the nature and origin of minerals in the preliminary discourse of the *Flore françoise* but his mentors evidently prevented him from doing so.[43] In 1780 he took on the formidable task of preparing the botanical section of the *Encyclopédie méthodique,* and until 1794, he published nothing but botanical studies.[44] To the public he must have appeared as a very competent professional botanist, no more and no less. Privately, his interests and goals were more comprehensive. While his technical work as a botantist was consolidating his position among the naturalists of the French scientific community, his more reflective tendencies were leading him in directions not fully expressed in his writings until the 1790s and afterwards.

A survey of the memoirs of the academy from the time of Lamarck's election in 1779 to the time of the body's abolishment in 1793 reveals just how much there was for a self-consciously independent and wide-ranging thinker like Lamarck to mull over during this period.[45] French science was enjoying a period of imsense vitality as Lamarck took his place at the academy among the likes of the chemists Lavoisier, Baumé, Berthollet, and Fourcroy; the astronomers Lalande and Lemonnier; the prominent students of mechanics Laplace and Coulomb; the experimental physicist Monge; the anatomists Daubenton, Portal, Vicq-d'Azyr, and Broussonet; the botanists Adanson, Jussieu, and Desfontaines; and the naturalists and mineralogists Desmarest, Sage, Haüy, and Tessier. (These scientists are identified by the sections of the academy they represented in 1785.) In botany, Lamarck's own field, the memoirs of the academy included not only accounts of useful species, previously unidentified species, and genera, and how to make a herbarium, but also discussions of plant growth, the ability of wheat grains to maintain their generative properties despite long periods of dormancy or exposure to extremes of heat and cold, motion in plants and animals, the response of plants to light, and the problem of whether the reproductive organs of plants truly exhibit the phenomenon of irritability. The writers on the natural history of animals described several new species and discussed the properties of an electric fish, the regeneration of lost parts in fish, respiration in fish, and the location of the tracheal artery in birds. The majority of the contributions in mathematics and astronomy probably meant little to Lamarck, but he must have taken interest in developments in other fields, such as Coulomb's announcement of his fundamental law of electricity, Haüy's discussions of crystallography, the various observations made on extraordinary meteorological phenomena, the report prepared by the special committee of the Academy investigating the highly controversial subject of "animal magnetism," and the special prize that the academy offered on one aspect of the way substances were circulated through the three kingdoms of nature by means of the interrelated processes of fermentation, putrefaction, combustion, vegetation, and animalization. He was clearly aware of the great excitement in the field of chemistry, where Lavoisier was leading the way to a new understanding of the phenomena of combustion, calcination, the compound nature of water, and so forth. Had Lamarck been interested only in living things, he would have had more than sufficient reason to take notice of the chemists' work, since their investigations included analyses of animal and vegetable substances and the study of chemical processes central to plant and animal physiology. As it was, the new developments in chemistry were of special interest to Lamarck because he had been elaborating a general chemical theory of his own.

Lamarck's own contributions to the memoirs of the academy between 1779 and 1793 were limited to three botanical papers: the first on a new genus of plants, the second on the "classes" of the plant kingdom, and the third on the flowers of the nutmeg tree.[46] He aspired, however, to broader accomplishments. For many fields of science, the barriers of specialization were just beginning to be raised. A single individual could still make a mark in several fields (a prime example of this was Lavoisier, who made major contributions to physiology and geology as well as to chemistry). In 1780, Lamarck presented a lengthy manuscript detailing his own "researches on the causes of the principal physical facts" to the secretary of the academy. To Lamarck's considerable disappointment, the committee assigned to review his manuscript never rendered an official judgment on it.[47]

At the academy, Lamarck was promoted from assistant to associate botanist in 1783, and from associate to full *pensionnaire* in 1790. In the 1780s and 1790s he also secured an affiliation with the outstanding institution of natural history of the time: the Jardin et Cabinet du Roi, headed by the illustrious Georges Louis Leclerc, Comte de Buffon. Buffon, through his own renown and the general excellence of the professors and demonstrators at the Jardin du Roi, through his direction of the internal affairs at the Jardin du Roi for nearly half a century, and through his cultivation of good relations with the high functionaries of the state (and his consequent ability to make major demands on the state's coffers to support his building projects at the jardin), had assured the jardin of a position of eminence in the scientific enterprise of the late eighteenth century.[48]

Having already extended his patronage to Lamarck for the publication of the *Flore françoise* and Lamarck's election to the academy,

Plan of the Jardin du Roi at its founding in the 1630s. By the time of Buffon's death in 1788, the grounds of the establishment had doubled in size and the number of buildings and greenhouses had tripled. From *Annales du Muséum d'Histoire Naturelle, 1* (1802).

Buffon in 1781 created for Lamarck the position of *correspondant* of the Jardin et Cabinet du Roi. This was not a salaried position. Its main purpose seems to have been to give Lamarck stature as he escorted Buffon's son, whom Buffon hoped would be the next *intendant* at the Jardin du Roi, on a scientific tour of Europe. Even so, the position represented an official connection with the Jardin du Roi, and as long as Lamarck still had the proceeds from the *Flore françoise* to draw upon, even an unsalaried affiliation with the important institution must have been welcome. By the end of the decade, however, Lamarck's financial resources were exhausted, and Buffon was dead. In 1789, Flahault de la Billarderie, who had supported Lamarck earlier, succeeded Buffon as intendant of the Jardin du Roi. La Billarderie proceeded to create for Lamarck a new position with a modest salary: botanist of the king and keeper of the herbaria of the Jardin du Roi.

Perhaps it is natural to ascribe one's own successes to merit while ascribing the successes of others to connections. This at least was Lamarck's perspective on the politics of the French scientific community when he found himself forced in 1789 to fight for the modest post at the Jardin du Roi he had only just received. His success in protecting his position is an indication that he was not only a talented naturalist but also a skillful lobbyist on his own behalf.

In 1789, as the French Revolution began, maintaining order and rescuing the government from financial disaster were the two major tasks before the National Assembly. The Committee of Finances, seeking economy measures wherever possible, perceived the Jardin et Cabinet du Roi as an establishment which had long failed to regard budgetary restraint as a virtue. In the course of considering how best to trim the establishment's budget, the suggestion was made that the care of the herbaria be the responsibility of the professor of botany, that the newly created position of keeper of the herbaria, occupied by Lamarck, be abolished, and that the position of assistant keeper of the cabinet of natural history, occupied by the geologist Faujas de Saint-Fond, likewise be suppressed. Lamarck quickly responded to this threat with two pamphlets, one in which he enumerated his own accomplishments, the other in which he argued that his position at the Jardin was an essential one that the professor of botany was too busy to handle.[49] The keeper of the herbaria, he insisted, was not "one of those useless positions, created under the old regime for the well-being of certain individuals in favor."[50] He presented himself as an individual who had not let himself be discouraged by obstacles placed in his path through the envies and personal preferences so characteristic of the politics of science of the previous days (he neglected to mention that he had not yet been effective in his new position because the botanists of the jardin, Desfontaines and Jussieu, had not permitted him access to the herbaria to rearrange them as he saw fit).[51]

The posts held by Lamarck and Faujas were not abolished. Instead, as the result of an address presented by the officers of the Jardin et Cabinet du Roi to the National Assembly, the Assembly agreed to let the officers draw up a plan of their own regarding the jardin's reorganization. Lamarck at this point offered a plan of his own to the Assembly.[52] This plan is a revealing one, for not only does it contain Lamarck's thoughts on the proper arrangement of natural history cabinets, it also serves as an index of the naturalists he esteemed most at the time the French Revolution began.

Having observed that many cabinets of natural history were "good for nothing" (since the only apparent motives for their arrangement were the desires to produce a pleasant visual effect or to flaunt the wealth of their owner) and that the cabinet of natural history at the Jardin du Roi would itself be improved by a more methodical arrangement of its collections, Lamarck indicated that no less than six naturalists would be necessary for putting the cabinet of the jardin in order: one to deal with the minerals, one to deal with the plants, and four to deal with the animal kingdom. Lamarck did not specify whether the courses in botany, chemistry, and anatomy that had been previously taught at the jardin would continue to be taught. Presumably he intended to leave that function unchanged. He did specify the positions and duties of the individuals responsible for arranging the collections, and he indicated the six persons he felt were best suited to the tasks: he himself for botany, Faujas de Saint-Fond for mineralogy, Daubenton for the quadrupeds and birds, Lacépède for the reptiles and fish, and two unnamed "persons of the highest merit" for the insects and for the "worms, lithophytes, and zoophytes."[53] Although he did not name these last two naturalists, from his comments it is evident that he was referring to his friends G. A. Olivier (for the insects) and J.-G. Bruguière (for the rest of the lower ranks of the animal kingdom). Daubenton's concerns as a naturalist have been sketched out above. In later chapters, in discussing the development of Lamarck's thought, it will be helpful to compare Lamarck's ideas with those of Daubenton and the other naturalists he wished to have as his colleagues.

The arrangement of specimens more for visual effect than for purposes of analysis, exemplified here by a plate from Dezallier d'Argenville's *Conchyliologie* (1780), was common in collectors' cabinets of the eighteenth century. Cabinets arranged in such fashion, Lamarck scornfully observed, contributed nothing to the progress of science.

Lamarck's personal plan for making the Cabinet du Roi a more useful institution was quickly superseded by a more comprehensive plan to reorganize the cabinet and jardin at the same time.[54] The officers of the Jardin et Cabinet du Roi, Lamarck included, presented this plan to the National Convention in the fall of 1790. It called for the foundation of a new institution: a museum of natural history. Sensitive to the utilitarian mood of the times, the drafters specified that the major purpose of the new institution would be to provide public instruction in natural history, with "natural history" understood broadly and "applied particularly to the advance of agriculture, commerce, and the arts." Appreciative also of the new spirit of egalitarianism, which coincided with their own self-interests, the planners saw the opportunity to do away with the all-powerful position of intendant, which Buffon had occupied with a free hand for so long at the old establishment. They proposed instead to elect a director from their own ranks who could serve no more than two years in succession. Twelve professorships, corresponding to twelve courses of instruction, were suggested for the new Muséum d'Histoire Naturelle. The subjects to be covered were mineralogy, general chemistry, chemical arts, botany at the museum, botany in the countryside, cultivation, ornithology and ichthyology, entomology and helminthology, human anatomy, comparative anatomy, geology and the instruction of naturalist-voyagers, and iconography. The number of professorships and the courses of instruction specified were not the result of a simple, objective appraisal of what would constitute a comprehensive program in natural history. There were twelve officers of the Jardin et Cabinet du Roi, all of whom intended to continue their work in the reorganized institution. Theirs were the names which appeared at the end of the plan of reorganization as the new professor-designates. With all but one exception they were cast in familiar roles. The exception was Lamarck.

The faculty proposed for the Muséum d'Histoire Naturelle in the plan of 1790 was a distinguished one. Eight of the men were members of the Académie des Sciences. The faculty was also relatively youthful, with the exception of Jean-Claude Mertrud, designated as professor of anatomy, and L.-J.-M. Daubenton, designated as professor of mineralogy, both of whom had been associated with the Jardin du Roi as early as the 1740s. The remaining ten members of the faculty were newcomers to the establishment, comparatively speaking, and ranged in age from thirty-four to forty-nine. Antoine-

François de Fourcroy, aged thirty-five, and René-Louiche Desfontaines, aged thirty-eight, were accomplished scientists and highly successful teachers. Fourcroy, who had been the professor of chemistry at the jardin since 1784, was an ardent exponent of both the new chemistry of Lavoisier and the revolutionary sentiments of contemporary politics. Desfontaines, who had been the professor of botany at the jardin since 1786, was a respected naturalist who had spent four years exploring the Barbary Coast. The elder botanist of the jardin, Antoine-Laurent de Jussieu, aged forty-two, had just brought to completion his great achievement in the natural classification of plants, *Genera Plantarum* (1789). Bernard de la Ville sur Illon, comte de Lacépède, the youngest of the group at thirty-four, had been the most recent of Buffon's collaborators on the *Histoire naturelle.* He was designated as "professor of the quadrupeds, birds, fish, etc." Antoine-Louis Brongniart, aged forty-eight, had served as the demonstrator in Fourcroy's course and was to be the professor of chemical arts. Antoine Portal, also forty-eight, was to continue in his old position of professor of anatomy, even though his active medical duties in the past had led him to neglect somewhat his responsibilities at the jardin. Barthélemy Faujas de Saint-Fond, forty-nine, who formerly had had the title of assistant keeper of the cabinet, was to be in charge of geology and the instruction of naturalist-voyagers. Gérard van Spaendonck, forty-four, was to continue as the institution's chief artist under the title of professor of iconography. André Thouin, forty-three, had distinguished himself through his activities as gardener-in-chief of the jardin, and was to have the new title of professor of cultivation. Finally, in the one case where an individual's past disciplinary affiliation was not maintained, J.-B. Lamarck, aged forty-six, was designated as "professor of the insects and worms." Lamarck had only a limited claim to expertise in this field, namely, he was an avid collector of shells and had assembled an impressive cabinet. He himself had just suggested that the two most talented naturalists in Paris in the fields of entomology and helminthology were Olivier and Bruguière, respectively. Lamarck, nonetheless, was to take on these fields as his own.

The National Assembly did not act upon the plan presented to them by the officers of the Jardin et Cabinet du Roi. In 1793, however, this same plan served as the basis of the proposal regarding the establishment of a Muséum d'Histoire Naturelle that Joseph Lakanal, a member of the Committee of Public Instruction, presented to the National Convention. The Convention, preoccupied with the Vendée rebellion, riots in several cities, and military advances by the Austrians, Prussians, and Spanish, passed Lakanal's proposal without discussion.[55] When the Muséum d'Histoire Naturelle opened its doors for instruction in 1794, the faculty it boasted was the same that had been named in the proposal of 1790, with the exception that Lacépède, who was being sheltered in the countryside because his life was in danger, was replaced by Étienne Geoffroy Saint-Hilaire, the youthful protégé of Daubenton. The officers of the Jardin et Cabinet du Roi, it should be noted, had proved successful as a group where the members of the academy had not. The officers of the jardin had been able to portray themselves as willing and useful citizens of the Republic who, rather than being favored members of the old regime, had themselves suffered under autocratic rule in the prerevolutionary period. They were rewarded with a new establishment which gave natural history a firmer institutional foundation than it had ever enjoyed before. By contrast, the academy was abolished.

While both the Académie des Sciences and the Jardin et Cabinet du Roi were attempting to weather the months of uncertainty in which any institution associated with the old regime was in jeopardy, new, independent societies sprang up dedicated to the pursuit and propagation of science.[56] Of these, one of the most significant was the Société d'Histoire Naturelle de Paris. During the 1790s, this society was a special center of activity for Parisian naturalists—especially those with revolutionary sympathies. It attempted to sustain the pursuit of natural history through the most difficult days of the Revolution, and if its own life was a precarious one which finally fizzled out at the end of the decade, it was still able in its day to claim such achievements as inspiring the National Assembly to sponsor a search for the lost expedition of La Peyrouse, offering regular courses of instruction in natural history in 1793 (no other institution did so at this time), offering prizes on important subjects in natural history, and publishing some valuable memoirs.[57] It erected a bust of Linnaeus at the Jardin du Roi (in 1790), it sent representatives to a celebration in honor of Jean-Jacques Rousseau (in 1791), and it attempted to have a voice in the organization of the Muséum d'Histoire Naturelle (in 1793).[58] Two problems plagued the society: it spent too many of its meetings debating political or organizational issues rather than issues in natural history and it was unable to sustain a successful scientific periodical.[59]

The bust of Linnaeus erected by the Société d'Histoire Naturelle under the cedar of Lebanon at the Jardin du Roi in 1790. When the bust was later overturned by vandals, some of the disciples of Linnaeus were convinced that the act had been encouraged by the opponents of Linnaeus' approach to natural history. From *Actes de la Société d'Histoire Naturelle de Paris* (1792).

The Société d'Histoire Naturelle de Paris came into being on the 27th of August, 1790, when eleven members of the Société Linnéenne de Paris gathered at the home of Louis Bosc, a Parisian naturalist, and agreed to rename their society. The former society, founded late in 1787, had held regular meetings through 1788 and numbered fifty-four resident members and forty-two correspondents by the end of that year. Thirteen members of the academy were among the resident members of the Société Linnéenne: Broussonet (who had the initial idea of establishing a society comparable to the Linnean Society of London), Daubenton, Desfontaines, Duhamel du Monceau, Fougeroux de Bondaroy, Fourcroy, Haüy, Lamarck, Lavoisier, Le Monnier, Malesherbes, Sage, and Thouin. But some members of the academy (including perhaps Lamarck himself) brought pressure to bear on the budding voluntary association because of its championing of the ideas of Linnaeus. This pressure, together with the general disruption caused by the political events of 1789, effectively cut short the vitality of the young society.[60]

The new Société d'Histoire Naturelle retained many of the members of the society that preceded it: twenty-two of the twenty-nine naturalists who belonged to the new society at the end of 1790 had belonged to the Société Linnéenne. During the next two years, as the membership of the Société d'Histoire Naturelle swelled to sixty-three, a dozen more members of the earlier society where admitted into the organization.[61] Prominent among those missing from the lists of the new society was Broussonet, who in addition to being a staunch Linnean was in political difficulty. Lamarck joined the Société d'Histoire Naturelle two weeks after it was founded.

The liveliest members of the new society in political matters were Fourcroy and Millin, both of whom were enthusiastic about the Revolution (Fourcroy, in October 1790, proposed the adoption of five articles for the society's rules regarding "the patriotic sentiments that members who compose it or are to be admitted must possess").[62] The busiest of the naturalists in terms of presentation of scientific papers was Bosc, at least in the early years of the society. Bosc bombarded his associates with observations on previously unidentified or insufficiently known organisms. Lamarck was a leader neither in political matters nor in presentation of scientific memoirs, but the records of the society show that he was an active member nonetheless. He advised explorer-naturalists on the kind of botanical observations they should make.[63] He excoriated Gmelin's new *Systema Naturae* as being "filled with the grossest errors"—

errors which he listed only in part because to have spent more effort on the book would have been a waste of his time.[64] He recommended candidates for membership, including Louis Reynier and Étienne Geoffroy Saint-Hilaire (it was Reynier who precipitated a political crisis in the society not long after his election by demanding that Ramond de Carbonnières be struck from the list of members because of his conservative political stands).[65] In 1792, with the aid of three members of the society—Olivier, Bruguière, and the chemist Pelletier—and with l'abbé Haüy, he was responsible for the publication of a scientific periodical, the *Journal d'histoire naturelle.* In the same year, he was one of several members who volunteered to give a free, public course in his specialty (which he designated as botany and conchology).[66] In 1793 he served as one of four commissioners in charge of planning a regular publication of the society's proceedings.[67] He was also designated as one of six directors of the society five years later, when it was in special need of revitalization.[68]

How ardent a republican Lamarck was is difficult to determine. Historians of science have cited him as a member of the Société de 1789, a society which, as its name suggests, endorsed the Revolution, but this proves to be a case of mistaken identity.[69] A letter of May 1790 from the French botanist Charles-Louis L'Héritier to the Englishman Joseph Banks does indicate that Lamarck and other Parisian botanists endorsed the Revolutionary cause: Jussieu was *lieutenant de maire,* Broussonet secretary of the municipality, Thouin representative of the commune, L'Héritier commander of a batallion, and Lamarck a private soldier.[70] In 1793 Lamarck undertook several tasks for the Committee of Public Instruction.[71] In 1794 he dedicated his *Recherches sur les causes des principaux faits physiques* "to the French people." In the dedication he asserted that he had resisted all those who had urged him to dedicate his *Flore françoise* to the minister, the king, or a certain nobleman, that he had never bowed down to anyone in the old regime, and that he wanted now to render homage to and be useful to "my kind, my brothers, my equals."[72]

Lamarck considered himself a man of principle. He was also a man with a sense of what was politically prudent. He had been a member of the nobility. It took a special dispensation from the Committee of Public Instruction to allow him to stay in Paris in April 1794, when a law was passed forbidding ex-nobles to reside in the city.[73] For a man with his background, a protestation of republican faith and demonstration of one's utility to the government was highly expedient. Lamarck, it is true, would later be associated with a theory of the nature and origin of life which was radical in its materialism. He never linked his scientific theory, however, to ideas of social reform. He does not appear to have been a political activist temperamentally. Instead, he seems to have been most at home in the quiet of the naturalist's cabinet.

Lamarck's expression of revolutionary sympathies did not hurt him in the first half of the 1790s. He received an indemnity of 3,000 livres in 1795 for past labors in the cause of science.[74] He pleaded in the same year that the loss of his pension at the abolished academy, the enormous rise in the cost of living, and the burdens of a large family had placed him in such a state of distress that he had neither the time nor the freedom of thought to cultivate the sciences fruitfully.[75] This plea was eventually answered with a small pension. Most importantly, he emerged from the early 1790s with a position in the French scientific community that was stronger than he had ever enjoyed previously. As of 1795 he was a professor at the Muséum d'Histoire Naturelle, a member of the first class of the Institut de France (the reconstituted academy), and an active participant in the leading scientific societies of the day: the Société d'Histoire Naturelle and the Société Philomathique (this last-mentioned organization having taken on a special significance in the fall of 1793 after the academy was abolished).[76] Indeed, though his scientific productivity did not peak until the first decade of the nineteenth century, by 1794 or 1795 his relative status within the French scientific community was at its height.

It is uncertain precisely when Lamarck's fortunes within the French scientific community began to slip. The records of the Société d'Histoire Naturelle for January 1795, however, have preserved an interesting juxtaposition of events that seem to foreshadow what was to come. Between the meeting of the Société d'Histoire Naturelle in which Étienne Geoffroy Saint-Hilaire proposed the young Georges Cuvier for membership and the meeting in which Cuvier was unanimously accepted, Lamarck read a report on his own new book, his *Recherches sur les causes des principaux faits physiques*.[77] In other words, at just the time Cuvier was beginning his meteoric rise in the French scientific community, Lamarck was embarking on a fruitless attempt to convince his contemporaries of the value of a system of physics, chemistry, and physiology that in fact showed little promise of advancing the frontiers of knowledge. While Cuvier's studies in comparative anatomy were effecting a major reform in the classification of the lower animals and providing a new rigor in the study of living and fossil vertebrates, Lamarck was fighting a rearguard action against the advances in chemistry achieved by Lavoisier and Lavoisier's followers. While Cuvier was establishing himself as a patient observer who claimed to go no further than the facts would allow him, Lamarck was gaining the image of a builder of groundless systems.

By the 1790s, there was already a long history of polemics regarding the proper role of facts and hypotheses in science. "System-building" had been identified early in the century as one of the greatest obstacles to scientific progress, and from Fontenelle to Condillac the "esprit de système" had been castigated.[78] Separating fact from theory was not considered difficult, at least not in principle. But counseling a greater attention to facts only sidestepped the major issue: how was one to know when enough observations had been made so that generalizations connecting diverse phenomena could be attempted successfully? Happy would be the man, Condillac suggested, who lived in a time that furnished him with enough facts so that he did not have to use his imagination.[79] But how was one to realize when that time had arrived? For a patient observer like Réaumur or Trembley this may not have been a burning issue. For a seeker of "les grandes vues," however, the problem was immediate. Buffon insisted that science have an experiential base, but he also wrote: "Let us put facts together to give us ideas."[80] In Lamarck's case, as in Buffon's, the urge to synthesize overrode contemporary advice to restrict oneself to discovering new facts.

Lamarck's desire to provide a grand synthesis of physical and chemical phenomena was not out of character. He continually conceived major undertakings for himself. Upon publishing his three-volume *Flore françoise* in 1779, he announced his intention to assemble a "universal" botany that would include the exact "analysis" and description of all the known plants.[81] Under a somewhat different format than he had originally intended, this is what his contribution to the *Encyclopédie méthodique* set out to be. His *Recherches sur les causes des principaux faits physiques* of 1794 was a two-volume effort, totaling more than 800 pages. In 1795, pressed for funds, he proposed to the Committee of Public Instruction that if paid in advance he would spend seven years on the preparation of a French "system of nature" that would present the "complete, concise, and methodical catalogue of every natural production observed up to the present." He told the committee that he had at first thought the work might be done by a society of naturalists but he convinced himself that then the whole work would have been "distorted, without unity of plan, without harmony of principles, and perhaps interminable in its composition."[82] The projects he entertained in the 1800s were no less grand than those he conceived as a younger man. In 1802 he announced his plans for his "terrestrial physics." Though only the *Hydrogéologie* part of this project appeared as initially advertized, the meteorological and biological sections did find expression under different formats by 1810. Between 1815 and 1822 he produced the work his contemporaries valued most, his seven-volume *Histoire naturelle des animaux sans vertèbres*. Fittingly, he permitted himself at the end of his career a final grand overview of human knowledge, his *Système analytique des connaissances positives de l'homme*.

Lamarck was not one to shrink from a study of details, but in many projects he saw his major contribution as one of providing a secure foundation for the science in question. As a naturalist, he was not content simply to identify and classify species. He saw himself as a naturalist-philosopher, a man whose thoughts centered on the broad processes by which nature brought about all its various results. While discussing the prospects of a science of meteorology in the early 1800s, he indicated that a real science was a body of knowledge that had "its philosophy, its principles, its consequences, and its problems to resolve." He regarded himself as the first to have "dared conceive the hope of setting forth the foundations of meteorology."[83] When he presented his biological views in 1802 he acknowledged that the novelty of these views called for a more extensive treatment than he had provided, but he supposed that the few people who had "seriously occupied themselves with the study of nature" could "easily supply the details and the applications that are missing here."[84] In the physiological part of his *Philosophie zoologique* (1809), he explained to his readers that he was not offering to them a whole new treatise on physiology (for "excellent works of this kind" were already available), but he was putting together "some well known general facts and fundamental truths on this subject, because I notice that their union sheds beams of light which have escaped those who have occupied themselves with the details of these things."[85] Earlier, in challenging the pneumatic chemists in the 1790s, his quarrel was not with their facts but with what he perceived to be their hasty raising of a superstructure without paying sufficient attention to the necessary foundations. Though not a chemist himself, he believed he could provide for the chemists the base on which to build a more permanent edifice. His physico-chemical speculations, however, may have reminded his contemporaries of Condillac's anecdote about a physician who built a chemical system and who, upon being told by a chemist that his facts were wrong, replied to the chemist: "very well, teach me [the facts] so that I can explain them."[86]

Lamarck's objections to what he considered to be the too narrow empiricism of contemporary sciences were expressed especially well in an unpublished manuscript of 1801 or 1802:

> It is at present a highly esteemed merit to occupy oneself only with the gathering of facts. One must search after them on all sides. One must consider them wholly in isolation. Finally one must confine oneself everywhere to the smallest details. This procedure alone is said to be estimable.
>
> As for me, I think it can be now useful to bring together the collected facts, and to strive to consider them as a whole, in order to obtain from them the most probable general results. Those who would conclude that in the study of nature we must always limit ourselves to amassing facts resemble an architect who would advise always cutting stones, preparing mortar, wood, iron-work, &c. and who would never dare to employ these materials to construct an edifice.[87]

Lamarck hated to be told that a major problem was out of reach of contemporary science, or that it could only be confronted after less fundamental issues were cleared up. What was one to do, he complained in 1802, with such important questions as whether or not the beds of the ocean had changed location in the course of the earth's history:

> Are we reduced to being able to form only arbitrary hypotheses, only gratuitous assumptions on these basic subjects? As many now think, must we avoid, under the pretext of this danger, envisaging the most important questions, only to occupy ourselves with the consideration of those of an inferior order, only to gather without end all the small facts that appear, and only to study them in isolation down to the most minute details without ever trying to discover the general facts or those of the first order, of which the others are only the last results?[88]

Lamarck was interested in facts, he said, but there were "big facts" and "small facts," and he was interested in the "big facts," the facts of the first order of importance.[89] Hypotheses were not to be foresworn as the bane of scientific progress. Instead they were to be recognized for what they were, and then used judiciously. Was one to remain silent with regard to certain facts rather than offer some plausible hypothesis to account for them, even if the hypothesis was not immediately provable? Lamarck did not think so. "I believe," he wrote, "that the course of silence is good for nothing. Every effort to lift the veil which hides nature's operations from us is useful. A mediocre idea often gives birth to a better one, and by force of trying one will perhaps obtain some successes. All that is important in such circumstances is to give as certain only that which is clearly demonstrated."[90] In principle, Lamarck's observations on the utility of hypotheses seem sound. It appears, however, that he was less adept at putting a hypothesis to a critical test than he was at devising hypotheses to cover what he perceived to be the "big" or general facts.

In the 1780s Lamarck turned to the Académie des Sciences for a judgment on his physico-chemical ideas, but the approval he sought was not forthcoming. With the suppression of the academy in 1793 he decided to present to the public the treatise on which the academy had not even deigned to prepare an official report. But the publication of his *Recherches sur les causes des principaux faits physiques* in 1794 was not a success. His views continued to be neglected. The general scientific community and the public at large did not receive his treatise with any more enthusiasm than had the academy's commissioners. Lamarck, for his part, was incapable of reevaluating the merits of his work. He regarded his *Recherches* as the consequence of careful thinking about the fundamentals on which the science of chemistry ought to be based. The neglect of his work, he concluded, was the result of a conspiracy against him engineered by persons who feared that their own theories (and hence their reputations) would be destroyed by his observations. "It is well enough known," he observed caustically in 1796, "that the interest of scientists is not always in accord with the interest of the sciences."[91]

Lamarck continued his attempts to gain a hearing for his physico-chemical ideas, but he remained unsuccessful. In a second physico-chemical treatise, *Réfutation de la théorie pneumatique* (1796), he took Fourcroy's recent *Philosophie chimique,* which he considered to be the "most complete, most precise, and best developed" presentation of the pneumatic theory, and showed point by point how his own theory could account for the facts Fourcroy cited.[92] The response to this attempt, however, was no more satisfying to Lamarck than the response to the first. A chemist in the city of Nancy bothered to criticize Lamarck's system, but the chemists of Paris did not.[93] Appalled by the continuing neglect of his views, Lamarck began in the fall of 1796 to present a series of physico-chemical memoirs at the meetings of the first class of the institute, hoping to provoke a discussion. Within a few months he gave up this new tactic in despair. He could not bear to read memoirs to colleagues who did not even pretend to disguise their boredom and annoyance with what he was doing.[94] He assembled these memoirs and published them in 1797 as his *Mémoires de physique et d'histoire naturelle.*[95] But he had not yet given up. He presented his new book to the Société Philomathique, accompanied by a special address in which he explained what he had been attempting to do. Refuse as one might the new considerations the book contained, Lamarck said, one could not deny it the merit "of having finally called the attention of enlightened men to the necessity of establishing bases of reasoning in physics . . . These fundamental bases," he insisted, "will serve to put a check on the excessive liking of certain men to build systems. . . . When one is not held back by invariable bases of reasoning, determined prior to all theory," he warned his auditors, "it is easy to fool oneself in considering the facts under certain aspects."[96] Once more, evidently, Lamarck was ignored. Three months after he presented his book of memoirs to the Société Philomathique he handed in his resignation to that organization.[97]

The whole experience was a painful one for Lamarck. He could not accept the idea that an activity was beyond his talents or expertise. He hated being ignored even more. As a youth his frustration was great when, as the youngest son in the family, he was destined by his parents for the clergy rather than the military. The "little abbé," as he was nicknamed, wanted a share in the glory that his older brothers reaped when they returned from military compaigns and great parties were held in their honor.[98] Later, as a botanist, he flew into a rage when a plant that he had given one name in his *Encyclopédie méthodique* was called by a different name in his presence.[99] Throughout his scientific career he displayed a keen concern with matters of priority.[100] To have labored for years on a project and then to have the fruit of his labors completely neglected was a situation he could not cope with. Rather than accept the possibility that his physico-chemical views lacked profundity, he meditated bitterly upon the means by which his views were being suppressed.

One did not have to be attacked frontally, Lamarck realized, to be destroyed by one's enemies. He knew all too well, he felt, what took place in scientific societies. In an unpublished manuscript of 1801 or 1802 he described the means he supposed were being used to destroy him. All that was necessary was a passing disparaging remark, a smile or a knowing wink, an air of disdain, a shifting of the discussion to another topic, and the uninitiated was apprised of the official judgment: Lamarck's physico-chemical system was not worth considering. And if someone without tact were to inquire directly about Lamarck's work, the agents of the established theory would be ready with a curt response: "you speak of a man who knows nothing, who is not acquainted with the facts, who has never made an experiment." By such means the conspirators ruined him behind his back and he had no means of defense.[101]

Lamarck's description of the methods by which his views were suppressed trails off with the following lament:

> History furnishes us with many sketches which teach us how much men of the first rank have had to suffer for having dared in good faith to seek the truth, instead of submitting to the dominant authority, instead of yielding to the accepted opinion.
>
> Let us compare here the ordinary savant, he who in his lifetime enjoys every advantage of a great reputation, and the rare, always unknown man who sincerely seeks the truth and prefers it to all other kinds of enjoyments. You will see in the first, beneath an exterior of modesty which he does not have, a conceited man, full of arrogance, bold, deciding all questions in a brief and curt manner, having always on his lips "that is demonstrated." This man is almost never found at home. One meets him in all the fine assemblies. He is a member of all the esteemed bodies, in charge of all the functions.[102]

Lamarck's words were not the words of a man who had always shunned the meetings of scientific societies. They were the words of a man who had fallen from the grace of the scientific establishment. It is not surprising to find that in the spring of 1797, when Lamarck was still attempting to gain a hearing for his physico-chemical views, his habit was to eat alone prior to the meetings of the Institute.[103] By the early 1800s he had concluded that he would find satisfaction only in the search for the truth, not in popular acclaim. Having failed to gain the approval of his contemporaries, he sought solace in his self-image as naturalist-philosopher. When he contrasted the man who rarely uses his intellect with the man who is a habitual thinker, his description of the latter was nothing less than a self-portrait:

> He is the man who, prepared initially by education, has contracted the useful habit of exercising the organ of his thought by devoting himself to the study of the basic knowledge available. He observes and compares all that he sees and all that affects him. He forgets himself in order to examine all that he can perceive. He varies without limit the acts of his intellect. He has gradually become accustomed to judge everything on his own, instead of adopting a blind confidence in the authority of others. Finally, stimulated by reverses and especially by injustice, he reascends peacefully through reflection to the causes which bring into existence everything we observe, whether in nature or in human society.[104]

CHAPTER TWO

The Background to Lamarck's Biological Thought

According to the French philosopher-historian Michel Foucault, the science of biology did not exist in the eighteenth century because *life* itself, as a concept, did not exist. What existed were living beings, which naturalists denominated by their visible characters.[1] Historians may debate the validity of this generalization, but it cannot be denied that the basic concern of the eighteenth-century naturalist differed from that of the nineteenth-century biologist. Indeed, neither the naturalist, the anatomist, nor the physiologist of the eighteenth century was particularly interested in the phenomenon of life as it manifested itself throughout the plant and the animal kingdoms. The naturalist concentrated upon outward and visible signs to distinguish living things from one another. The anatomist and physiologist concentrated upon structures and functions, but only of the human body. Thus when Cuvier, in 1800, defined the comparative anatomist's task as comparing all the species and pursuing "life and the phenomena that compose it in all the beings which have received some parcel of it," he was signaling, if not carrying off singlehandedly, a major transformation in scientific thought.[2] Correspondingly, when Lamarck coined the word "biologie" in 1802, he was giving a name to a basically new subject, not one that had existed and had been a primary interest of his since his career began in the 1770s. Early in his career Lamarck distinguished the study of the "laws of vegetation" from the study of the visible characters of plants ("their color, their size, their duration, and in general all which allows us to distinguish them one from another"), and it was the second subject which he said belonged to botany, "properly speaking."[3] His *Flore françoise* was an analysis of the visible characters of plants, not a study of plant physiology. Two decades later, the focus of his attention had shifted away from the external characters of living things to their internal organization and to the vital faculties their organization bestowed upon them. He was becoming a biologist, though he had not started out as one.

The lecture in which Lamarck first introduced his evolutionary views, the opening discourse he delivered to his students at the Muséum d'Histoire Naturelle in the spring of 1800, provides a convenient introduction to the central themes of his thought on the eve of his conception of the new science of biology.[4] "If it is true," he told his listeners, "that in order to study natural history in a profitable manner . . . it is first necessary to embrace with the imagination the vast ensemble of nature's productions, to raise oneself high enough by this means to dominate the masses of which this ensemble appears to be composed, to compare them among each other, [and] finally to recognize the principle traits which characterize them; . . . [then] I must begin by recalling succinctly for you the major distinctions that nature herself seems to have established within the immense variety of her productions, the path [*marche*] or order she seems to have followed in forming them, and the singular relations she has caused to exist between the ease or difficulty of their multiplication and their particular nature."[5] In part Lamarck was only describing the traditional task of the naturalist. But he did not regard himself as an ordinary naturalist. He considered himself a naturalist-philosopher, and as such he intended to tell his students about (1) the basic differences between the inorganic and organic realms, and, within the organic realm, between the plants and the animals; (2) the natural relations among nature's productions, as evidenced by the gradation of complexity of organization displayed in both the plant and the animal scales (Lamarck's new insight as of 1800 was a phylogenetic interpretation of these scales); and (3) the relation between complexity of organization and possession of special faculties, with special reference to the different reproductive potentials of organisms of different complexity (and to the balance of nature by which organisms of great fecundity are kept in check). Each of these topics was of immediate importance to Lamarck in the formulation of his new view of organic change.

Each would also play a part in shaping his new science of biology. None of them was basically new to him at the turn of the century, however. His recent studies in invertebrate zoology had provided him important new perspectives on all three (especially the third), but he had already met these topics, made some decisions on them, and in some measure preconditioned his later thinking about them in his previous years as a botanist and as an elaborator of a speculative physiology. Though it is inappropriate to call Lamarck a biologist in the early years of his career, some of the basic assumptions of his later biological thought were formed prior to his becoming professor at the museum in 1793. The present chapter will deal with the early period of Lamarck's thinking—the period up through the publication of his physico-chemical treatise of 1794—and with the foundations of his biological thought that were laid during that period.

THE "NATURAL RELATIONS" AMONG LIVING THINGS

In 1783 Lamarck identified six different aspects of botany that deserved the attention of the serious student of that science: (1) the principles of plant physiology; (2) the natural relations among the plants; (3) the means of identifying plants; (4) plant nomenclature; (5) the history of botany; and (6) plant cultivation.[6] Plant physiology, he suggested, could be looked upon as the base of all the others. Plant identification was the area in which he had registered his first major success. The subject that had come to attract him most, however, was that of the "natural relations" (*rapports naturels*) among the plants. This was a part of botany, he stated, that could be appreciated only by the "philosopher-naturalist."[7] In his view, the knowledge of the natural relations of the plants had to be regarded as "the true philosophy of botany and the end that one proposes for himself when he dedicates himself entirely to the cultivation of this fine science."[8] Later, when he turned his attention to the study of the invertebrates, this concern with natural relations continued to inspire and direct his thinking. As he told his students in 1799, beyond paying attention to the invertebrates for reasons of health or economics, "the consideration so attractive for the naturalist, so worthy of a philosopher-observer," was the study of natural beings according to "their true relations."[9]

Lamarck's early concern with natural relations must not be construed as an endorsement of the idea of evolution. The relations in which he and his fellow naturalists were interested were of a formal, not a phylogenetic, nature. The goal of the major French botanists of the latter half of the eighteenth century was to discover the "natural order," an arrangement in which each plant would be placed nearest those plants that resembled it most closely in its most essential characteristics.[10] Linnaeus had spoken of the natural method as the ultimate goal of botany, but he had based his own classification of the plants on a "sexual system" that was admittedly artificial rather than natural.[11] By relying almost exclusively upon a consideration of the stamens and pistils of plants, he created groupings that occasionally included plants differing markedly from each other except for their generative organs. Critical of this system, the major French botanists (most notably Bernard de Jussieu, Adanson, A.-L. de Jussieu, and Lamarck) sought to determine the natural relations of the plants not by the organs of fructification alone but by all the different plant characters (only Adanson went so far as to suggest, however, that in determining plant affinities all the different characters should be weighted equally). Buffon had scoffed at the botanists' idea of natural method, likening it to the alchemists' idea of the philosopher's stone.[12] A.-L. de Jussieu used the same analogy, but Jussieu did so to stress the desirability of the goal and the difficulties standing in the way of its attainment, not the possibility that the natural order might prove to be illusory.[13] Jussieu indeed maintained that "botanists and zoologists . . . should put aside their systems and work together in the search for this order," for the existence of this order was "proven" and the order had been shown to be "founded upon invariable principles."[14]

Lamarck's early thoughts on the "true relations" among nature's productions are perhaps best analyzed by considering first his understanding of the basic differences between "brute" and "organized" bodies and between plants and animals and then by examining his view of the natural order as it existed within each of the three traditional kingdoms of nature. He always insisted that fundamental differences exist between the inorganic and the organic realms of nature and, within the organic realm, between the plants and the animals. He also clung tenaciously to the idea that the natural order within each of nature's kingdoms was best represented by a scale of graded complexity.

From his first publication, the *Flore françoise* (1779), Lamarck drew a common but nonetheless fundamental distinction between inorganic and organic beings. Inorganic beings, he said, "grow by the juxtaposition of substances . . . , and not by the effect of any internal principle of development"[15] while organic beings "are provided with organs suitable for different functions, and possess a very marked vital principle and the faculty of reproducing their own kind."[16] A few years later, in defining the word "species" (*espèce*), he was careful not to apply this word to the mineral kingdom because of the fundamental difference between minerals on the one hand and plants and animals on the other: "Without the constant reproduction of similar individuals, there cannot be real *species* [*espèces*]. That is why it is wrong to qualify as species the diverse kinds [*sortes*] of minerals that have been observed."[17]

For Lamarck, the crucial differences between inorganic and organic beings had important implications for the proper arrangement of nature's productions. He commented in 1794 that for "nearly twenty years" he had wanted to publish his thoughts on "how unjustified the naturalists were who regarded as possible the formation of an uninterrupted chain that would comprise all the beings in nature."[18] His criticism of the proponents of this idea spilled forth in one long, breathtaking stream:

> As if there were some relation between a being endowed with life, susceptible of development, of a state of vigor, and then of decline; a being subjected necessarily to losses, which at the same time has the faculty of making up for these and maintaining itself; a being which produces its kind, and which exists itself only because it was similarly produced by another individual of its species; a being finally the life of which is essentially subjected to constant limits, and which submits to an inevitable death at the end of the prescribed term of its duration; as if, I say, such a being could be compared with a piece of mineral, that is to say with a being which is not an individual, which is by no means endowed with life, which has no nutritive faculty in itself, which is never produced by its kind, and the duration of which would be unlimited, like that of a piece of gold, if nature did not tend to subject it to a conjunction of circumstances suitable for its alteration; in a word, a being which having no real principle of life, cannot be subject to death.[19]

Other naturalists, rather than accepting as fundamental the division of nature's productions into three kingdoms, had also divided nature's productions into brute, unorganized bodies on the one hand and living, organized bodies on the other. Daubenton, Pierre-Simon Pallas, Johann Blumenbach, Félix Vicq-d'Azyr and A.-L. de Jussieu all insisted upon this distinction.[20] At the Muséum d'Histoire Naturelle in the 1790s Desfontaines was presenting the idea to students in botany at the same time Lamarck was presenting it to students in invertebrate zoology.[21] Lamarck had company in his claim that there is an "infinite distance" between living and inorganic beings.

Somewhat fewer were those who believed, with Lamarck, that no transitional forms connected the plant and the animal kingdoms. The subject of "zoophytes"—organisms presumed to be part animal and part plant—had attracted the attention of naturalists throughout the second half of the century. Proponents of the idea of a single "chain of being" linking all of nature's productions naturally regarded the discovery of such organisms as freshwater polyps and sea anemones as the confirmation of the postulated transition between the animal and the plant kingdoms. Buffon, perceiving no absolutely essential difference between the animals and plants, suggested that the freshwater polyp could be considered "the last of the animals and the first of the plants."[22] Others, without accepting the idea that the plants and animals formed a kind of chain, did believe that the organic kingdoms were united by transitional forms. Both Pallas and Vicq-d'Azyr, for example, believed in the existence of organisms that were half-plant, half-animal.[23] Daubenton denied that any transitional forms linked the animals and the plants, but he went on to question whether certain simple living forms should be included among the plants or animals at all. It seemed to him that some new major categories might have to be established.[24]

For Lamarck, the distinction between plants and animals was clear and did not necessitate the formation of any new groups. When, in 1786, he presented a table of the different classes of living organisms, he arranged the plants and animals in two distinct series and classed the zoophytes with the animals.[25] In 1794, going beyond the common distinction that animals are capable of sensation and voluntary motion while plants are not, he identified the exclusive feature of plants to be "the very remarkable property of combining together free elements and being the primary cause of all the compounds that exist on our globe."[26] No animal, he observed, is capable of nourishing itself with elementary substances. All animals have to depend directly or indirectly upon plants for their food. Only plants, by means of their own vital activity in conjunction with sunlight, can form compounds. Later, he decided that the basic property that distinguished animals from plants was the property of irritability, which he traced to a chemical difference between animal and plant substances.[27] Whatever distinguishing criterion he was stressing at a particular point in his career, however, he always insisted that a major hiatus separated the animals and the plants.

In the course of his lectures and writings Lamarck sometimes referred to the "immense chain of natural beings." This was never, however, an endorsement of the concept of the chain of being in its broadest sense. Even though Lamarck's work as a systematist was dominated by the assumption that a natural arrangement of the productions within each of nature's kingdoms would be basically linear, and even though he acknowledged the existence of fine gradations in the animal, plant, and mineral series taken separately, he consistently denied that these different series could be joined. In this respect, and in others, he showed himself a confirmed opponent of Charles Bonnet, the major exponent of the chain of being in the latter part of the eighteenth century.[28]

Those who contemplated the possibility of the natural method that would place the different organisms in their natural order had different expectations as to the general form this natural order would take. Linnaeus suggested that the arrangement of plants according to their natural affinities would be comparable to the arrangement of territories on a geological map.[29] Donati, the naturalist of the flora and fauna of the Adriatic, suggested that natural affinities would be best represented by a network in which each knot was attached to several threads and the threads were joined together or were crossed over one another in countless different ways.[30] A.-L. de Jussieu spoke instead of "the chain of beings" and of "rising by an imperceptible gradation from the least perfect animal to the most perfect, from the simplest herb to the largest plant."[31]

Like Jussieu, Lamarck, in his *Flore françoise,* portrayed the natural order as a chain of graduated complexity. Perhaps Lamarck owed this view of the natural order to Haüy, for he acknowledged Haüy had helped him with his discussion of the subject.[32] Whatever the origin of Lamarck's views on the subject, his belief that the natural order was basically linear came to be one of the most prominent features of his thinking. The historian Henri Daudin, in two masterful analyses of the idea of serial arrangement in natural history, has indicated how this idea functioned first in Lamarck's botanical and then in his zoological studies.[33] In Daudin's words, the idea of serial arrangement was the "master doctrine" on which Lamarck most insisted, "the active formula of his thought in natural history from the beginning to the end of his career."[34]

Significantly, Lamarck's belief in a hierarchy of beings was not based primarily on metaphysical considerations. He was not concerned with Leibniz's principle of sufficient reason, nor did he make any principle of plentitude or continuity a cornerstone of his thinking. At times he described the natural world as "full."[35] He supposed that within each organic realm there was a graded series of complexity of organization. But, as indicated above, his view on the distinctions between the organic and the inorganic realms and between the plants and the animals clearly set him apart from such upholders of the chain of being concept as Bonnet. Unlike Bonnet, who exalted both the "law of continuity" and the philosopher who expounded it, Lamarck concentrated his attention upon nature's productions and processes and did not bother with metaphysical arguments for why nature's operations should be what they are.[36]

Lamarck's thought on the major distinctions among nature's productions prevented him from accepting the common aphorism "natura non facit saltus,"[37] but while he rejected the idea of transitional forms between the major divisions of nature, he endorsed the idea of gradations between forms within each division. Buffon had used the idea of the gradation of nature's productions to attack the reality of the systematic categories presented by Linnaeus.[38] Lamarck did likewise. Lamarck argued that the view of nature's operations accepted by Linnaeus himself contradicted Linnaeus' notion of the reality of genera:

> Linnaeus, as well as many others, has . . . said in his works that nature makes no leaps, which signifies, if I am not mistaken, that the series of her productions must be finely graded throughout its length. Now, this single consideration destroys the possibility of finding the whole of nature's productions divided by her in a number of special groups well distinguished from one another, such as the genera ought to be. The limits of each of these groups would be precisely the leaps that it is recognized nature does not make.[39]

Interestingly enough, Lamarck did not carry this argument denying the reality of the genera and the higher taxa to the logical conclusion of denying the reality of species as well. The trend of his philosophical reasoning, it would seem, was not always in perfect harmony with his work as a naturalist.

Lamarck also used his view of gradation within the plant kingdom to deny the utility of Jussieu's principle of the subordination of characters. The desire "to divide and subdivide by groups with the aid of a supposed subordination of clear and striking characters," Lamarck maintained, was always thwarted by the consideration of the true relations among the members of the various divisions. The distinct groupings implied by the principle simply did not exist in nature.[40] Later Lamarck was more charitable toward the principle of the subordination of characters, and he himself divided both the plant and the animal scales into classes that he apparently regarded as natural. However, he continued to have difficulty reconciling the belief that nature was graded and continuous with the supposition that nature's productions could be arranged in distinct, natural groups.

At first, in the *Flore françoise,* Lamarck expressed the hope that the natural order would be a linear arrangement of the different plant species.[41] Previous attempts at ordering the plants had not been entirely successful, he claimed, because botanists had wanted to force nature "to arrange her productions as a general arranges his army: by brigades, by regiments, by batallions, by companies, etc." (Linnaeus had divided the plant kingdom into classes, orders, genera, species, and varieties and had pointed out the military analogy himself).[42] These divisions were foiled, according to Lamarck, by "the admirably graded nuances that nature has established among the majority of the plants." "Nature," he said, "offers for our attention and our speculations an immense collection of beings, among which each species is distinguished from the others by a perceptible and a constant difference; and the gradation of these differences is the foundation of the order we propose." Lamarck sought "a whole submitted to fixed rules," but rules that would "tend only to determine the place each species must occupy in the general series" and would not cut the whole up into distinct divisions.[43]

To begin setting up the natural order, Lamarck indicated, the botanist had to be able to answer at least one of two questions:

Which is the plant which appears to us most alive, the best organized, in a word, the most perfect?

Which is the plant which we must judge naturally the least complete in its organs, and which seems most removed from the other plants by its different aspects?[44]

Judging that the second question was easier to answer than the first, Lamarck indicated that it was among the cryptogams, among the tiny organisms that could be considered the "first sketches" of plant life, that the series should begin. At this point, before he had come to view the plant and the animal series in phylogenetic terms, he supposed that the series beginning with the simplest plant would have to be reversed to represent the *true* natural order. The true natural order would begin with the most complex plant and gradually descend from there to the simplest.[45]

Lamarck did not achieve a serial arrangement of the species in his *Flore françoise*. He presented there only a "sample" of his natural order, employing genera rather than species and admitting that the rank given to each of the genera had been determined in too vague a manner. Having acknowledged these deficiencies, he expressed the hope that his rough arrangement would give an idea of what the true natural order must be like.[46]

By 1785 Lamarck had evidently despaired of arranging the species, genera, or even the orders in a single series. He offered instead six classes that seemed to him to represent the "true order of gradation" in the plant kingdom.[47] Within the serially arranged classes he also presented the families in series, but he acknowledged that in too many cases "the order of the families still appears arbitrary."[48] His general operating principle remained the same as before. He put at the opposite ends of the scale the two plants most unlike each other in complexity of organization: "Although a *Byssus* [a lichenized algae] and a pear be two beings of the same kingdom, and both true plants, the difference in their organization is so considerable, that it indicates that in the general series of beings of this nature these two plants must be proportionally separated from one another."[49] In setting forth the six plant classes Lamarck noted "with pleasure" that his arrangement of the plants formed "a perfect counterpart with the large sections that divide the animal kingdom."[50] Later, when he came to study the zoology of the invertebrates more closely, he increased the number of animal classes but did not give up the idea of a general hierarchical arrangement for each of the organic kingdoms. He maintained firmly that at least the different "masses" within each kingdom could be arranged serially according to the relative complexity of their respective organizations.

Just why Lamarck should have clung so tightly to this idea of serial arrangement is not readily explicable, though as of 1802, if

Lamarck's table of 1785 displaying both the "true order of gradation" in the plant kingdom and the way his arrangement of the plants formed "a perfect counterpart with the large sections that divide the animal kingdom."

* *Etres organiques vivans, aſſujettis à la mort, & qui ont la faculté de ſe reproduire eux-mêmes.*

Animaux.	Végétaux.
1. *LES QUADRUPEDES.*	*LES POLYPÉTALÉES.* 1.
1. Terreſtres onguiculés.	Thalamiflores. 1.
2. Terreſtres ongulés.	Caliciflores. 2.
3. Marins.	Fructiflores. 3.
2. . . . . *LES OISEAUX.*	*LES MONOPÉTALÉES,* . . . . 2.
1. Terreſtres.	Fructiflores. 1.
2. Aquatiques à cuiſſes nues.	Caliciflores. 2.
5. Aquatiques nageants.	Thalamiflcres. 3.
3. . . . . . *LES AMPHIBIES.*	*LES COMPOSÉES.* 3
1. Tétrapodes.	Diſtinctes. 1
	Tubuleuſes. 2.
2. Apodes.	Ligulaires. 3.
4. . . . . *LES POISSONS.*	*LES INCOMPLETTES.* . . . . 4.
1. Cartilagineux.	Thalamiflores. 1.
	Caliciflores. 2.
2. Epineux.	Diclynes. 3.
	Gynandres. 4.
5 . . . . *LES INSECTES.*	*LES UNILOBÉES.* . . . . 5.
1. Tétraptères.	Fructiflores. 1.
2. Diptères.	Thalamiflores. 2.
3. Aptères.	
6. . . . . *LES VERS.*	*LES CRYPTOGAMES.* . . . . 6.
1. Nuds.	Epiphylloſpermes. 1.
2. Teſtacés.	Urnigères. 2.
3. Lithophytes.	Membraneuſes. 3.
4. Zoophytes.	Fongueuſes. 4.

Botanique. Tome II.

not a year or two sooner, he did have in his theory of evolution a physical explanation of why a general series should exist in each organic kingdom. His early adherence to the idea lacked any obvious metaphysical base or the support of an evolutionary theory. What he had instead was the example of many naturalists before him who had found the order of decreasing complexity an obvious way to classify things.[51] In comparing more familiar forms with less familiar forms, the choice of the human type as the basic unit of reference was inevitable. Other forms were then arranged according to how closely they approximated this most highly organized type. By subtracting more and more features from the human model, one arrived eventually at the least complex forms of animal organization. Lamarck, whose career was characterized by a desire to grasp nature's fundamental laws in their simplicity, found the idea of serial arrangement quite congenial. This arrangement, he acknowledged, might not be a perfect representation of nature's true plan, but it was well suited to human intellectual capabilities:

It is first of all certain that we will never grasp the vast and magnificent plan that directed the Supreme Being in the formation of this Universe . . . The plan of Nature embraces at once the immensity of the whole and that of its details. It consists in the relations that an infinite Wisdom has arranged as much between the exterior and interior qualities of each individual, and the destination of this individual considered either in itself or with regard to the whole Universe to which it is tied by an infinity of threads of which the majority are imperceptible to us.

In the absence of this knowledge which will always be denied to us, we must cling to that which is more adjusted to our intellectual abilities, and limit our researches to arranging individuals relative to our way of seeing and comparing objects.[52]

Later, in a manuscript entitled "On what remains to be done to give to botany the degree of perfection it cannot do without," Lamarck wrote:

As for the whole series of plants, I will say first that I do not know if nature has really formed a well-developed chain among the plants, just as I am perfectly unaware of whether she has established a similar one among the animals. I know well that the idea of a single chain continuous through the beings of the three kingdoms of nature is entirely chimerical and does not have the least foundation. . .

Whatever the case, whether nature has really formed a continuous chain for the plants . . . it seems to me always very necessary in facilitating the progress of botany to work at the formation of the most natural series possible.[53]

Near the end of his career, Lamarck came to believe that at least two distinct series were involved in the order of production of the invertebrates, but he still thought, at least for pedagogical reasons, that a simple series ought to be used "in our works and in our courses to characterize, distinguish, and make known the observed animals."[54]

Lamarck maintained his enthusiasm for arranging nature's productions linearly despite all those who supposed that the natural order would be of another form—the reticular arrangement or the geographical map analogy being the two major alternatives to the graded series.[55] In the latter part of the 1790s—the period when Lamarck's evolutionary ideas began to take shape—the linear arrangement of nature's productions was under attack. Though many of these attacks were directed primarily against the ideas of Bonnet, Lamarck's thoughts on serial arrangement were, at least on one occasion, also subjected to criticism.[56] At the turn of the century, as Lamarck saw it, a number of authors had come to look upon the reticular scheme of classification as "sublime."[57] In his own work on the classification of the invertebrates, however, Lamarck was finding to his satisfaction that it was easier to do for the animals what he had long been trying to do for the plants: arrange them in a single series of increasing (or decreasing) complexity. His supposition that the natural order was a scale of graded complexity had a decided effect upon the way his evolutionary theory originated and developed.

THE NATURE OF LIFE

Thus far out discussion of Lamarck's early thought has focused on Lamarck's ideas as a naturalist, ideas that would later be central to his understanding of the production of organic diversity but which did not, prior to the late 1790s, bear so directly on his understanding of the nature of life. Questions regarding the nature of life and the consequences of organic activity, were, however, questions that interested Lamarck from early in his career. Through 1794 his general comments on the nature of life appear neither especially incisive nor suggestive of things to come. In his long physico-chemical treatise of 1794 he represented life as an incomprehensible vital principle. But if this general view seems unpromising for his later evolutionary theorizing, the same cannot be said of some of his specific statements about vital phenomena. His ideas on organization, irritability, and the role of subtle fluids in vital processes, though all fairly commonplace for the eighteenth century, played major roles in the evolutionary theory he eventually developed. His idea that vital activity is the ultimate source of all chemical compounds, an idea that was not commonplace for the time, provided Lamarck with a model of change in the inorganic world that was strikingly similar to the model of change he ultimately proposed for the organic world.

In his earliest writings Lamarck spoke of life as an animating or vital principle, though he seems to have felt that plant life could be understood in terms which were basically mechanistic.[58] By 1794 he was more dogmatic in his statements on the special character of life. One of the things man would never understand, he asserted, was the cause of "the existence of organic beings and that which constitutes the life and essence of these beings, since matter with all its properties seems to me to be in no way capable of producing a single being of this nature."[59] Living things possess a "particular principle," he maintained, "the origin and essence of which can no doubt not be assigned physically."[60] But when Lamarck said that the essence of life would forever remain a mystery to man, he was not recommending an end to all physiological study. He still supposed that the physiologist could investigate "the physical cause that maintains life in living things, gives rise to their development, and finally produces their death."[61] Though matter could not produce life, it was nonetheless true that life could not exist without matter, and the physiologist could at least study the material conditions of life's existence. The incomprehensible principle which is life, he explained, "resides essentially in a particular movement of the organs of beings . . . , a movement that transmits and perpetuates itself through the generations, but which can exist in each individual only under certain circumstances and during a necessarily limited time."[62] He supposed that while the principle of life was not to be understood simply by the arrangement of organs, this principle manifested different faculties depending upon the organization and chemical composition of living bodies.[63]

The living body, as Lamarck perceived it, was "a whole composed of diverse sorts of parts, some more or less solid and the others fluid."[64] All compounds, he also supposed, have a natural tendency to decompose. Organisms, being chemically complex, were necessarily subject to the tendency to decompose, but this was counterbalanced in them by the principle of life. Matter was assimilated in the body as the result of the principle of life or "organic movement."

Assimilation, in Lamarck's view, was the key to understanding the growth, maturation, and inevitable death of living things. The young organism grew, Lamarck explained, because it assimilated more matter than the ever-present processes of decomposition caused it to lose. The suppleness of the young organism's fibers was what made this possible. The relatively slight resistance that "organic movement" encountered within the body accounted for the special facility with which the organism incorporated foreign material into its own substance. The incorporation of matter into its own substance, however, caused the organism's fibers to lose their flexibility. In the natural tendency of compounds to decompose, as Lamarck saw it, the "elastic principles" always escaped more quickly than the "fixed" or "earthy principles," leaving the fibers of the body more solid. With this increase in solidity, the organism reached its period of greatest strength. At this point, the quantity of matter assimilated and the quantity of matter lost were equal, and growth ceased. Then, as the organism's fibers continued to retain more fixed than elastic principles, the organic movement responsible for assimilation met with increasing resistance and was unable to make up for all the losses the body suffered. The body became increasingly rigid. Death was the inevitable result of the process of life.[65]

The idea of life as a movement of fluid and solid parts of the organism, a movement that in itself leads necessarily to its own cessation, was standard in the eighteenth century. It can be found in the physiological treatises of Hermann Boerhaave, Friedrich Hoffmann, and Albrecht von Haller.[66] It can be found in the *Encyclopédie* and the writings of Charles Bonnet.[67] Buffon, like the others mentioned, described old age as a period of increasing ossification.[68] Hoffmann wrote in his *Medicina rationalis systematica:* "We learn by careful observation that motion is the cause of all bodily changes, that in motion also lies the basis of life and health; that the very causes of diseases act upon the solid and fluid parts of our body in no other way than through motion; nor do therapeutic agents exert their effect except by motion."[69] Von Haller introduced his *First Lines of*

Physiology with the observation that the human body is composed of solids and fluids, and he concluded his treatise with a chapter on nutrition, growth, life, and death: "We feel the beginnings of decay even in youth itself... Even in that blooming season the solid elements of the body are augmented, the chinks through which the humors flow are lessened, small vessels filled up, and the great attraction of the cellular texture has added a density to the whole body."[70] It was clear to Von Haller that "when those causes continue to operate by rendering the matter of the body more dense, by diminishing its irritability, and augmenting the quantity of earth, it is not possible but decrepit old age must succeed."[71]

Lamarck believed he could add to what the great physiologists of the eighteenth century had shown, by developing his idea that the tendency of all chemical compounds is to decompose. Boerhaave, according to Lamarck, had assumed that health resided primarily in the state of the body fibers, while Hoffmann had emphasized how health depended on the state of the circulation. Not wishing to demigrate the ideas of these men, Lamarck suggested they were both right, but they had mistaken derivative causes of health for the principle cause: the balance between the "force of assimilation" and the "effectuation of the tendency to decomposition."[72] The phenomenon of animal heat, Lamarck suggested, could also be accounted for by the tendency of compounds to decompose. Rather than suppose, as Boerhaave had done, that animal heat was to be explained by the friction between the fluid and solid parts of the body, Lamarck supposed that animal heat was produced by the continual disengagement of "fixed fire" from the blood. This disengagement was most prompt, and animal heat was consequently highest, in those animals that displayed the most "organic action."[73]

The differences between the physiological system proposed by Lamarck and the systems proposed by Boerhaave, Hoffmann, and other eighteenth-century physiologists need not be detailed here. What should be noted is that Lamarck approached physiology as a system-builder, and in his belief that all compounds tend to decompose he found a unifying explanation for a broad range of physiological phenomena. At the same time, he incorporated in his thinking much that was commonplace in his day. When he dropped his view of life as a vital principle, and when he came to believe that life in the simplest animals depended upon external stimuli rather than an incomprehensible vital force, he continued to think of life in terms of organic motion. Life remained, in his view, essentially a function of the contained fluids and the containing solids of the organized body. And in the mechanical effects of fluids moving within the living body, he found that he had an explanation not only of the phenomena of growth and development but also of organic mutability.

The ideas on vital activity and organization that Lamarck held prior to the mid-1790s do not readily identify him as an emerging biologist. They show him operating instead in the more traditional roles of naturalist and armchair physiologist. He was interested in complexity of organization as an index of an organism's place in the natural order. He paid less attention to the faculties an organism's organization allowed it to have. He did concern himself with some physiological issues in his physico-chemical treaties of 1794, but he concentrated on *human* physiology. He justified this focus with the observation that man was the being with not only the most organs but the "most perfect" organs as well. Though he indicated that each species had faculties that depended upon the number and perfection of its organs, he did not elaborate on the idea.[74]

Lamarck's general inattention to the relation between faculties and organization up to 1794 is not surprising, given that most of his time was devoted to botanical studies. Plants do not display a wide variety of faculties. Plants, as he defined them in his *Flore françoise,* were those organic beings "that develop and live, but [do so] without being endowed with any sensation, and without having movements other than those caused by the organism's own organization or the action of external bodies."[75] Animals, in contrast, in addition to developing and being alive, were capable of sensation and spontaneous movement. In plants, Lamarck maintained, the quality of being alive was "the effect of organization alone."[76] In animals the quality of being alive depended upon a "principle of sensation."

If Lamarck's botanical work provided him little basis for exploring the relation between complexity of organization and possession of vital faculties, it did, however, provide him with a model of vital activity that he later generalized. In attributing motion, and hence life, in plants to nothing more than the plant's organization and the external influences upon it, he was presenting an idea that later proved basic to his understanding of the nature of life in the simplest animals, and, as he came to argue it, these were the animals that had to be studied if the essence of life was to be truly identified.

Lamarck did not intend the *Flore françoise* to be a text on plant physiology. The long section on "principles of botany" in that work was devoted almost exclusively to the definition of terms used to designate plant characters. In some brief comments on the motion of fluids in plants, however, he gave some indication of what he meant when he said that in plants the quality of being alive "is the effect of organization alone." Sap rose in the plant during the day, he explained, as the result of heat dilating the plant's tubes. In the day, the pores in the leaves served as excretory ducts, exhaling the more fluid part of the nutritive juices of the sap. At night, when the temperature dropped, the parts of the nutritive juices that had not been exhaled were deposited in the interstices of the plant fibers, there to be assimilated. At the same time, the pores in the leaves took on the function of receiving atmospheric "juices," which descended all the way down to the roots. Lamarck had little to say about the importance of light for plant growth. He simply noted that light favored plant development, and that this was perhaps due to light's "analogy or even identify with electrical matter." Electricity, he explained, "accelerates the course of liquids, and consequently must augment the flow of the nutritive juices and hasten the progress of the vegetation."[77] In sum, for plants as well as animals, Lamarck understood life to be largely a matter of hydraulics. The annual plant died, he supposed, when the first frosts of the winter produced a tightening of its tubes, which prevented the motion of its sap. The human being died when its fibers became so rigid as to resist the "organic movement" that constituted the "principle of life."[78]

As indicated, Lamarck posited in the *Flore françoise* a major distinction between plants and animals: life in the latter depended not just upon organization but also upon a "principle of sensation." To maintain this distinction, he had to account for a phenomenon that seemed to lessen the gap between the plant and animal worlds: the phenomenon of "sensitive" plants, or plants that contract when touched. Lamarck did not broach this problem in the *Flore françoise,* but the demands of alphabetical order forced it quickly upon him when he began work on the "botanical dictionary" that was his contribution to the *Encyclopédie méthodique,* for the "acacia" constituted a genus of plants that was notable for the "singular and marked irritability" of some of its species.[79]

Lamarck did not believe that anyone had explained plant irritability satisfactorily, so he proceeded to give his own thoughts on the subject in discussing the "common sensitive plant," *Mimosa pudica.* He supposed that elastic and subtle fluids disengaged themselves continually and abundantly from plants, especially in warm weather. Such losses occurred in all living things, he presumed, but in the case of the sensitive plant these elastic and subtle fluids were not all exhaled as soon as they were formed. Instead they were amassed in the plant, giving to the plant's most mobile parts "a kind of tension and rigidity that maintains them in the state of extension one sees when [the plant] is open." When the plant was touched, the subtle fluids within it were dissipated and the resultant void caused the plant to contract upon itself. If his explanation of the phenomenon was correct, Lamarck noted, then irritability or "sensitivity" in plants was not due to any real sensation on the part of the plant but instead to a "purely mechanical cause."[80]

Lamarck had more to say on the subject of sensitive plants. When in 1785 he divided the plant kingdom into six, serially arranged classes, he noted that it was "almost uniquely" within the class of the greatest complexity, the *Polypetala,* that the phenomenon of irritability was exhibited (*Mimosa pudica, Oxalis sensitiva,* and some other well-known sensitives were in this class). It was as if, he remarked, "the principle of life renders itself more manifest in these plants and brings them near in a way to the other organic beings in which irritability is joined to a more perfect quality called sensibility."[81]

Lamarck did not suppose that irritability and sensibility were the same thing. Sensibility, he indicated in 1789, was directly related to complexity of organization. It was strongest in the "perfect animals"—"those in which animalization is the most complete in all its faculties." It became weaker and appeared even to disappear altogether as one moved from the most perfect animals to the least perfect animals. Irritability, in contrast, seemed to grow in intensity as one reached the "last orders" of the animal kingdom. The "less perfect" an animal was, the more extended was its faculty of irritability. Thus, although the muscles of a man or quadruped would cease to exhibit irritability within an hour and a half of the organism's death, the viscera of a frog would still be irritable twenty hours after removal from the frog's body (Lamarck was present when this was demonstrated in one of Vicq-d'Azyr's courses). Irritability, Lamarck observed, was found in a number of plants, but this irritability, he continued to maintain, was, in apparent contradistinction to irritability in animals, due to "an entirely mechanical cause." As before, he explained plant irritability by the action of a subtle fluid that collected in the "utricles" of plant tissue and then escaped when the plant was touched.[82]

Lamarck's discussion of plant irritability was not strikingly original. The distinction he drew between irritability and sensibility had been made a generation earlier by von Haller in his dissertation "on the sensible and irritable parts of animals." Von Haller called a part of the body "irritable" if it became shorter when touched and "sensible" if it occasioned signs of pain or discomfort in the experimental animal (or if, in the case of the human being, the impression of being touched was transmitted to the soul).[83] Experimentation alone, he argued, could determine which parts of the body were sensible and which were irritable. The source of sensibility, he maintained, was the nerves. Irritability he assumed to be "a property of the animal gluten," which probably depended on "the arrangement of the ultimate particles."[84]

Lamarck's use of subtle fluids in his physiological explanations was no more exceptional than his use of von Haller's distinction between sensible and irritable parts of the body. To explain the phenomena of heat, electricity, and magnetism, theorists in the eighteenth century postulated the existence of a variety of subtle, elastic, and unweighable fluids.[85] Not surprisingly, such fluids were also pressed into service to explain the vital phenomena of irritability, sensation, and growth. For example, when Friedrich Hoffmann sought to explain the power of the heart and arteries, he did not feel compelled to call upon the mind, the sensitive soul, or some vital force. Instead, he suggested that the "adequate, constant, and proximal cause of the vital motions of systole and diastole, by which all natural actions take place" was the result of "a very delicate fluid, warm, elastic, contained in the finest tubules of the membranes and nerves and in the blood itself, which is the cause of their motions, and of life, health, and diseases."[86] When l'Abbé Nollet, the famous French experimentalist and impressario, discovered that water would flow faster from a capillary tube that was electrified than from one that was not, it occurred to him that electricity "might have a remarkable effect on organized bodies, since organized bodies could in some fashion be regarded as "hydraulic machines."[87] Nollet suspected that the action of electricity "might well make itself felt on the sap of plants, or give to the fluids that play a role in the animal economy some movement which would be advantageous or harmful to them."[88] He proceeded to electrify

Testing whether subjection to the prolonged action of electricity has an influence on the growth of plants and the transpiration of animals. From Nollet, *Recherches sur les causes particulière des phénomènes électriques* (1749).

plant seeds and discovered that this stimulated their generation and growth. Believing that electricity would increase transpiration in animals (the pores of an animal's skin could be considered "the extremities of an infinity of very fine tubes") and that transpiration involved the exhalation of a fluid, he concluded that prolonged exposure to electricity ought to cause an animal to lose weight. He tested this on cats, pigeons, buntings, chaffinches, sparrows, and even flies. The experimental results confirmed his conjecture. He found that the electrified animals lost more weight through transpiration than did the controls. The effect, moreover, was proportionately greater in small animals than large ones. He supposed that electricity might have a therapeutic value in treating cases of paralysis, and he and a number of eighteenth-century physicians attempted to discover whether this was so.[89]

At the time Lamarck began his career, the conventional wisdom concerning the role of the subtle fluids in the production of vital phenomena was that of Joseph Toaldo, the professor of astronomy, geography, and meteorology at the University of Padua. Toaldo won a prize from the Académie des Sciences in Paris in 1774 for the best essay on the question: "What is the influence of atmospheric phenomena on vegetation? And what practical consequences can be drawn, relative to this subject, from the different meteorological observations made up to now?"[90] In his essay Toaldo commented on the way daily temperature changes cause the motion of plant fluids. The terms he used were virtually the same as those Lamarck employed a few years later in his *Flore françoise*. Toaldo also commented on the way the "electric fire" stimulated plant growth, perhaps playing an even greater role in this respect than heat and humidity. In the first place, electricity affected living things indirectly by producing rainstorms and other atmospheric phenomena that benefited vegetable development. In the second place, electricity affected living things directly by "penetrating and agitating the fluids and solids of all living bodies" and by "exciting especially the circulation of fluids in the small channels or capillary tubes of plants, as well as sensible and insensible transpiration, which the good or bad condition of plants and animals depends upon."[91] During stormy weather, Toaldo remarked, when the humid vapors of the air absorbed considerable amounts of electric fire, animals and especially birds were "very agitated, sometimes sad, sometimes lively, according to their acquisition or loss of this fire which animates them. Even plants show visible signs of external change through the alteration of their system."[92]

On into the 1790s meteorologists measured the amount of electricity in the atmosphere, as well as the temperature and humidity, confident that all these factors influenced living things. The ex-academician Quatremère Disjonval, losing his sanity in a prison cell in Holland, convinced himself that spiders and his own "spermatic vessels" were unusually sensitive to atmospheric electricity. The web-spinning behavior of spiders and the activity of his sex organs, he announced, were better prognosticators of shifts in the weather than the mercury barometer.[93] Lamarck credited spiders with being able to sense changes in the weather twenty or even thirty hours before the changes occurred, but he never mentioned Quatremère's professed talents in this realm.[94]

As subtle fluids were called upon to perform various heroic services, scientists like Lavoisier warned their contemporaries that "it is with the things we can neither see nor feel that it is especially important to guard against flights of the imagination."[95] Lavoisier and his fellow academicians Bailly, Franklin, Le Roy, and de Bory prepared the official academy report debunking the fluid that Mesmer and his followers claimed was responsible for the phenomena of "animal magnetism."[96] But Lavoisier did not fail to believe in the existence of certain subtle fluids. Indeed it was in setting forth the evidence for the existence of an "exquisitely elastic" and subtle fluid that he offered the cautionary observation just cited. The subtle fluid in question was the "matter of heat," which he had named "caloric." He described this as "a real and material substance, or very subtle fluid, which, insinuating itself between the molecules of all bodies separates them from each other."[97] This was the fluid responsible for all phenomena of heat, including the heat of animals.

The role of subtle fluids in plant and animal life was discussed throughout the 1780s and 1790s. There were debates concerning the reality of the effects of electricity on plant growth and the legitimacy of Galvani's analogy between the animal body and the Leyden jar.[98] There was disagreement over the cause of animal heat (Lamarck found that Lavoisier's idea that respiration was a kind of combustion was too extraordinary to be true, but both men explained animal heat in terms of subtle fluids—"fire in the state of expansion" in Lamarck's case, "caloric" in Lavoisier's).[99] But the many distinctions made in the eighteenth century with respect to the identification, the action, and the interrelations of the subtle fluids need not be detailed further here. Suffice it to say that Lamarck drew upon a wealth of experimentation and speculation by others in formulating his own thoughts on the role of subtle fluids in nature. When, in the early years of the nineteenth century, he accounted for a host of biological phenomena by the action of such fluids, he was in large measure simply calling upon the dilative and accelerating properties these fluids had long been assumed to have.

IN LAMARCK'S EARLY discussions of life, the two most prominent ideas are the idea of life as organic motion and the idea that vital activity is the sole cause not only of organic complexity but of chemical complexity as well. Though the first of these ideas was characteristic of physiological thinking in the eighteenth century, only a few of Lamarck's contemporaries sympathized with the second. Lamarck first announced his ideas on mineral origins in 1786. He claimed "that the minerals are all true products of the successive alterations that the remains of organic beings experience through time."[100] This view was not entirely novel. It had been announced over a decade earlier by the chemist Antoine Baumé, a member of the academy. Lamarck acknowledged the similarity of his position with that of Baumé. Rather than relying on Baumé's authority, however, he backed up his thoughts on the origin of minerals with his theory that the tendency of all compounds is to decompose and with observations that he made in the mines of Saxony and Hungary during his travels in 1781–1782.

Lamarck held a traditional view of fire, air, water, and earth as the four basic elementary principles. He denied that these elements or principles had any "affinity" for each other. His assumption was that each had a tendency to separate itself from any combination in which it was found. In all the mines he had visited, he claimed, he found "that the soil newly formed at the surface of the earth by the remains of organic beings was more complex, softer, and less dense, and that in proportion as one went down into the earth and penetrated a more anciently formed soil, this soil, altered and changed in the course of time, was consistently harder, denser, less complex, and always more and more quartziferous and vitreous."[101] Lamarck believed that dead organic matter passed through successive mineral stages as the different elementary principles in the original organic substance slowly disengaged themselves at different rates and in different quantities. The earthy part of the organic substance, masked initially by the other principles present, revealed itself more and more as the other principles escaped, until what was once an or-

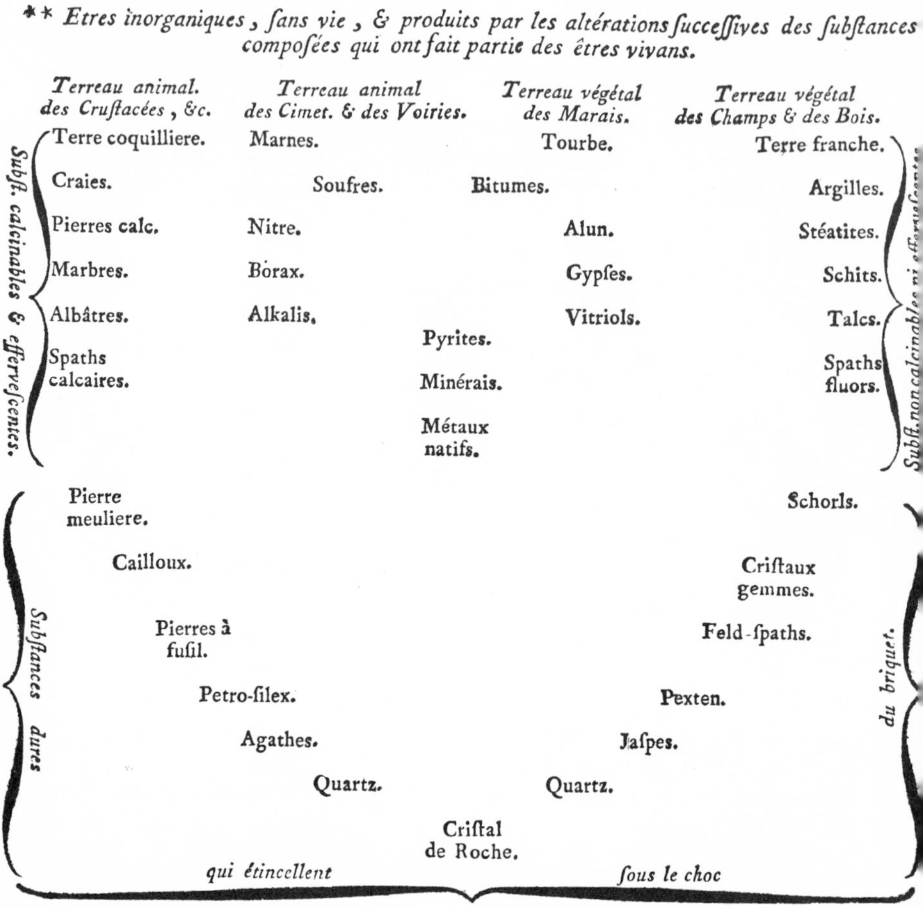

*** Etres inorganiques, ſans vie, & produits par les altérations ſucceſſives des ſubſtances compoſées qui ont fait partie des êtres vivans.*

Terreau animal. des Cruſtacées, &c.	*Terreau animal des Cimet. & des Voiries.*	*Terreau végétal des Marais.*	*Terreau végétal des Champs & des Bois.*

Subſt. calcinables & efferveſcentes.

Terre coquilliere.
Craies.
Pierres calc.
Marbres.
Albâtres.
Spaths calcaires.

Marnes.
Soufres.
Nitre.
Borax.
Alkalis.

Bitumes.
Pyrites.
Minérais.
Métaux natifs.

Tourbe.
Alun.
Gypſes.
Vitriols.

Terre franche.
Argilles.
Stéatites.
Schits.
Talcs.
Spaths fluors.

Subſt. non calcinables ni efferveſcentes

Subſtances dures

Pierre meuliere.
Cailloux.
Pierres à fuſil.
Petro-ſilex.
Agathes.
Quartz.

Criſtal de Roche.

Quartz.
Jaſpes.
Pexten.
Feld-ſpaths.
Criſtaux gemmes.
Schorls.

du briquet.

qui étincellent ſous le choc

Lamarck's table of 1786 illustrating his idea that the different minerals are formed through the successive breakdown of organic matter.

ganic substance became quartz and then finally pure rock crystal. Plant remains gave rise to one series of minerals, and animal remains gave rise to another. In each instance, however, the end product was the same.[102]

Lamarck's theory of mineral origins was not identical to Baumé's. Lamarck had a more restrictive view of the means by which compounds could be produced. Baumé maintained that nature is unable to form compounds directly from the four primitive elements. He maintained also that "the plant is the instrument Nature uses to combine the elements in the first place, and to form together with the animals all the combustible matter that exists in nature."[103] He asserted, however, that once plants had combined the elements to produce combustible matter, and once a multitude of animals had transformed "vitrifiable earth" into "calcareous earth," then nature could alter *and combine* these products to produce all the different compounds in existence.[104] Lamarck's claim was that nature could form no compounds whatsoever. Nature could combine neither the basic elements nor the primary products of organic activity. After the products of organic activity were formed, all the other compounds that appeared in nature were solely the result of the gradual process of decomposition.[105]

What is most striking about Lamarck's theory of mineral origins is its similarity to his later explanation of organic diversity. Lamarck himself never commented upon this similarity, but the two theories seem to be variants of a single pattern of thought. In each case, Lamarck stressed the gradual and successive production of forms. His view of how the different organic beings developed gradually from the very simplest forms of life seems to be basically the inverse of his view of how the different minerals had their origin in the gradual decomposition of organic products.

Until Lamarck developed his ideas on life and organic change, there appears to have been a fundamental incongruity in his thinking. The origin and essence of one part of nature's productions, the minerals, could be easily explained, while the origin and essence of organized, living bodies were completely inaccessible to the human intellect. There is nothing to indicate, however, that Lamarck regarded this situation as incongruous, at least not in the early 1790s. Profoundly impressed by the great differences between living and nonliving things, he could not have assumed blithely that what was true of the inorganic realm was also true of the organic realm. But conscious or not of what appears to have been a major incongruity in his early thinking, he in effect resolved it by doing for the plants and animals what he had done already for the minerals—explaining their origins as the result of natural processes operating slowly over long periods of time.

THE BALANCES OF NATURE

In the late 1790s, when contemplating the nature of life in the simplest animals and the apparently inverse relationship between reproductive potential and complexity of organization, Lamarck was inevitably drawn to consider how the populations of simple and extraordinarily fecund forms of life were kept in check. This was not a subject on which he had had much to say prior to the 1790s. In the *Flore françoise* he had remarked that the Creator had provided for the conservation of the species and that "diverse obstacles" restrained nature's reproductive capacities within "proper bounds."[106] In an article on forests for the *Encyclopédie méthodique* he had emphasized the importance of organic detritus for soil fertility and warned that the activities of man could upset some natural balances.[107] But in his early writings he never dwelt on the topic of "nature's balances." By the end of the century, however, this topic had become a critical issue for him.

The existence of "balances" in nature was cited in a number of the natural theologies of the eighteenth century as evidence of the wisdom and goodness of the Creator.[108] Early in the century, William Derham had devoted a section of his *Physico-Theology* to a discussion of "the balance of animals, or their due proportion wherewith the world is stocked," observing that "the whole surface of our Globe can afford Room and Support only to such a number of all sorts of Creatures. And if by their doubling, trebling, or any other Multiplication of their Kind, they should increase to double or treble that number, they must starve or devour one another. The keeping therefore the Balance even is manifestly a Work of the Divine Wisdom and Providence."[109]

Linnaeus, perhaps the keenest eighteenth century observer of the harmonious interrelations among nature's productions was likewise sensitive to the theological implications of these interrelations, though the implications did not dominate his writing on the subject. In the *Systema naturae* Linnaeus described the "polity of nature" as follows: "On account of the plants the herbivores exist; on account of the herbivores the carnivores, and of these latter, the large ones for the small ones; and man (as an animal) for the large ones and for everything, though principally for himself; so that by the necessarily destructive and oligarchical domination of one over another, the proportion and equilibrium of natural things is manifested, with the splendor of the republic of Nature."[110] Elsewhere he defined the "economy of nature" as the "wise disposition of natural substances, established by the master of the universe and according to which all things tend to a common purpose and have reciprocal uses."[111] He cited Derham's view on the correspondence between the amounts of food available and the number of beings in existence,[112] and he gave his own illustration of a food chain: "The *aphid* lives on plants. The *Musca aphidivora* eats the aphids. The hornet-fly (*Asilus*) lays snares for these. The *Libellulae* or dragonflies feed upon the hornet-flies. The spiders hunt the dragonflies. The sparrows take the spiders in flight, and they fall into the talons of the birds of prey."[113] Linnaeus' understanding of the economy of nature, summed up in his own words, was that "in order that the continuous series of beings be lasting, God in his wisdom willed that everything with breath be sensitive to the care of perpetuating its own species, and that the natural substances lend themselves to a mutual aid, and finally that death or the destruction of one be the principle of life or generation of another."[114]

Other eighteenth-century writers also referred to the interrelations that Derham had called the "balance" and Linnaeus had called the "economy" of nature. Buffon supposed that some species had probably become extinct, but he still felt that in the ordinary course of nature the fecundity of each species was kept in check: though there were periodic oscillations in the number of individuals in a species, over time these oscillations tended to balance each other out.[115] As he put it in 1765, "Nature does not protect man and the major species at the expense of the most trifling species or vice versa. She sustains them all."[116] Bonnet expressed the same general idea in his *Contemplation de la Nature:* "There are eternal wars among the animals, but things have been so wisely arranged that the destruction of some brings about the conservation of others and the fecundity of the species is always proportional to the dangers which menace the individuals."[117]

Bonnet spoke of "eternal wars among the animals" and Linnaeus spoke of "a continual and reciprocal struggle of beings,"[118] but the idea of competition between species—to say nothing of competition between individuals of the same species—was not the dominant theme in eighteenth-century discussions of the balance of nature. The naturalists of the period supposed that each species had been allotted what was necessary for maintaining its existence, no more and no less. They typically assumed that the relations between species exhibited a kind of wisdom on the part of nature or the Creator and that the balance or economy of nature was such that no species could be lost as the result of natural processes. In short, they assumed that the general order of nature was always preserved. Lamarck shared this view with the majority of his predecessors and contemporaries. At the very end of the century, he was confronted with the idea that a whole host of species once living had become extinct. He found this new idea unacceptable. It was inconceivable to him, given the general balance of nature and the uniformitarian character of nature's processes, that a broad-scale extinction of species could ever have occurred.

The idea of natural balances appealed to Lamarck. It fit well his notion of a universe governed by law. He found balances that conserved the species, balances that were responsible for the health of the individual, and even balances that preserved soil fertility. Thus, just as he believed that good health depended on new matter being assimilated as quickly as old matter was lost, he believed that soil fertility depended on a continuous supply of organic detritus counteracting the natural processes of soil decomposition and erosion. He noted in particular that when a country was deprived of all its forests, the destructive forces outweighed the restorative forces: "The soil of the country [deprived of its forests] loses little by little all its suppleness, its softness; its particles become more and more divided, no longer adhering to one another." Soon the particles were unable to retain moisture, and eventually they reached "the state of a sand that becomes more and more vitreous."[119]

Forests, Lamarck maintained, were valuable to the soil in several respects. They protected the soil from the sun and the wind and thus kept the soil from becoming too arid. They enriched the soil through the detritus of the plants and animals living there. They also influenced the weather: a countryside broken up by forests and woods received more rain than a countryside that was simply a treeless plain. Clearing all the forests of a country was disastrous: "We regard as a certain principle that a country where the forests and woods have been entirely destroyed, in order to occupy the terrain with particular cultivations which are annually productive, loses insensibly its whole fertility, and must one day reach a sterility capable of making man and all the other living beings there abandon it."[120] In this instance, at least, the balance of nature was not so flexible that man could not upset it.[121]

LAMARK'S EARLY THOUGHTS on the distinctions among nature's productions, the natural method, the nature of life, the relations between organization and faculties, and the balances of nature all reveal a thinker who was firmly rooted in the latter half of the eighteenth century. Lamarck still managed, nonetheless, to distinguish his thinking from that of his contemporaries. In some instances he took a common notion, such as the mechanical effects of moving fluids within the living body, or the influence of the environment upon living things, or the contribution of organic detritus to the earth's soil, and then developed it in an original, and generally unpopular, way. In other instances he took an equally common idea, such as the linear arrangement of nature's productions or the four-element theory of chemistry, and clung to it after others had given it up. Like many a systematic thinker of the time, he believed in a well-ordered universe directed by laws, and he had faith that these laws, in the last analysis, were simple. Despite an occasional acknowledgment that man could not know the Creator's mind, or that man had to make do with his own limited powers of understanding, a belief in the comprehensibility of the universe underlay and sustained his activities as a naturalist-philosopher. Though in 1794 he maintained that the most basic questions about the nature and origin of life lay beyond man's grasp, his confidence in his own reasoning ability and in the universe's ultimate rationality made it not so difficult for him to extend the sphere of his inquiries to these questions. And once he had given more attention to the "animals without vertebrae," once he had studied the striking degradation of organization between the most complex of these animals and the most simple, once he had considered what it meant for the simplest of these animals to be living, it became possible for him to conceive of a science of biology that would treat the nature, the origin, the faculties, and the development of life.

CHAPTER THREE

Eighteenth-Century Views of Organic Mutability

In the *Philosophie zoologique* (1809), his most famous biological exposition, Lamarck presented his view of the origin of species as his own conclusion and only indirectly suggested that the common assumption of species immutability had ever been questioned. The conclusion drawn by "nearly everyone," he wrote, was that every animal has a fixed organization and structure that are and always have been invariable.[1] Lamarck was indeed original in his idea that nature successively produces the different forms of animals one from another, beginning with the simplest and moving gradually to the most complex. He was also accurate in claiming that the fixity of species was widely accepted. But that is not to say that the issue of organic mutability had not been raised in the eighteenth century. The issue confronted naturalists in a variety of contexts. It asserted itself when they sought to account for new forms that they believed had not existed previously. It appeared again when they considered the effects of acclimatization on species transported from one place to another. It became even more significant as they developed a new perspective on the earth's age and history. The notion of organic change also occurred to thinkers who were not naturalists and who had concerns of a more broadly philosophical nature, such as affirming the fullness of the Creation and the continuity of nature's productions or working out a materialistic explanation of the universe. Lamarck was a professional naturalist. His involvement with the concerns of the naturalist is evidenced in the theory of organic change he eventually developed. In the boldness of his intuitions and assertions, however, and in his concern with the nature of life rather than just the distinguishing characters of the different living things, he often resembled the *philosophe* more than the naturalist. Though he was capable of devoting painstaking attention to the characterization of plant and animal species, he was also capable of making extraordinary intellectual leaps. A prime example of the latter is afforded by his statement that "once the difficult step [of admitting spontaneous generation] is made, no important obstacle stands in the way of our being able to recognize the origin and order of the different productions of nature."[2] The statement is not reminiscent of Lamarck's fellow systematists. It is reminiscent of Diderot.

If there was a standard assumption about the origin of species in the eighteenth century, it was that the species of the present were identical in kind and number with the species of the original Creation. This idea was endorsed in 1736 by the great Swedish naturalist Linnaeus in the proposition "there are as many species as the Creator produced diverse forms in the beginning."[3] Linnaeus himself, however, soon called this idea into question, and by 1769 the French botanist Michel Adanson was able to write: "One of the most celebrated and discussed questions in the past several years in natural history and especially in botany is to know if the species among the plants are constant, or if they change; that is, if by sexual reproduction or otherwise, new species are formed which become set in their own right, reproducing themselves constantly under this new form without reverting to that of the plants from which they originated."[4]

Linnaeus dropped the idea that the number of species had been constant from the Creation when he was confronted, early in the 1740s, with what appeared to be the direct origin of one species from another.[5] A plant was called to his attention which resembled the plant *Linaria* in all respects except for its floral structure. Linnaeus named the new plant *Peloria* and wrote a dissertation upon it. When he sent a copy of his paper to von Haller in Göttingen, he stated: "I beg of you not to suppose it [*Peloria*] anything else than the offspring of (*Antirrhinum*) *Linaria,* which plant I well know. This new plant propagates itself by its own seed, and is therefore a new species, not existing from the beginning of the world; it is a new genus, never in being till now. It is a mule species in the vegetable kingdom, propagating itself by transmutation of one plant into another."[6]

Earlier in the century, James Marchant, a French botanist, had suggested that the Creator might have created only generic types in the beginning and that the various species arose later from these original types.[7] Linnaeus adopted a similar view. Though he came to realize that *Peloria* was not the distinct plant he had presumed, the case of *Peloria* led him to contemplate the phenomenon of hybridization, and he soon discovered what seemed to him to be a number of indisputable instances of the production of new species by means of hybridization.[8] He eventually developed the view that God, at the Creation, had not created all the different species directly but instead had created only one species for each of the different natural orders. These species, being mutually fertile, hybridized to form the different genera, and further hybridizations produced the diverse species found at the present.[9]

The idea that new species could arise through hybridization was adopted by Adanson, who, in his *Familles des plantes* (1763), acknowledged explicitly that "species change."[10] After reviewing the writings of Marchant, Linnaeus, and Gmelin on the appearance of new plant forms, Adanson suggested additional mechanisms of organic change. Changes similar to those effected by hybridization between different species, he supposed, could be effected either by "the reciprocal fertilization of two individuals different in some way, though of the same species; or by cultivation, the terrain, the climate, the dryness, the humidity, the shade, the sun, etc. These changes are more or less prompt, more or less durable, disappearing with each generation or perpetuating themselves for several generations, according to the number, strength, and duration of the causes which combined to form them in the first place."[11]

The effects of cultivation upon plant forms were well known, Adanson claimed. Plants transported from one climate to another or taken from the wild state and cultivated in gardens differed so much from their original forms that "even the most well-trained botanist [could] scarcely recognize them."[12] The varieties of cabbage, for example, though so different in their appearance, had been shown by Morison to have all been produced from one another.[13]

The idea of the origin of new species appealed enough to Adanson that he found in it an explanation for the discrepancies between the botanical knowledge of the ancients and that of the eighteenth century. The naturalists of antiquity, he suggested, had neither described their specimens badly nor failed to identify vast numbers of species existing in their day. If plants the ancients described were nowhere to be found in the eighteenth century, it was simply because these plants had changed form.[14]

The appearance of new plant forms in modern times was also attested to by Antoine-Nicolas Duchesne, who, like his father, was a horticulturalist of the king at the garden of Versailles. Duchesne found in 1763 a race of strawberry that had not existed previously. The new race had simple leaves rather than the ordinary palmate ones with three divisions. This discovery led Duchesne to reflect on the problems of defining genus, species, race, and variety, and of determining the natural characters of species.[15] One of his primary conclusions was that the term *race,* which Buffon had employed in natural history of animals, should be introduced into botany. Otherwise, forms that were only races would be designated incorrectly as species, species would be mistakenly called genera, and so forth.[16] This sort of confusion, Duchesne explained, had led some naturalists to the conclusion that neither species nor genera were stable in the plant kingdom.[17] He proceeded to argue that the evidence cited by Linnaeus and Adanson for the existence of fertile hybrids was inconclusive. He concluded that species were fixed and immutable but that new races could be formed when the "accidents" responsible for individual variation were pronounced enough to be passed on from one generation to the next.[18] With Buffon, he assumed that the "mongrels" arising from the union of different races of the same species might become the founders of new races, but that the hybrids produced by the mating of individuals of different species could not reproduce themselves.[19]

Duchesne had documented the appearance of a new kind of plant. He supposed, however, that the change he had recorded was one that had taken place within the limits of the conventional species. He wrote of arranging the races of strawberries in a "genealogical tree," maintaining that "the genealogical order is the only one nature indicates, the only order that fully satisfies the mind."[20] But he only applied this thinking to the analysis of intraspecific differences. When he sought to represent the affinities between different species, he did not use a "genealogical tree" but rather a table that enumerated similarities conceived in a purely formal sense.[21]

The botanists of the eighteenth century did not fail to consider the idea of organic change. Once the question was raised, however, the general consensus was that significant organic change—that is, the actual transformation of species—was not part of nature's plan. In 1769, three years after the publication of Duchesne's *Histoire naturelle des fraisiers,* Adanson brought before the Académie des Sciences a reconsideration of the question of whether new plant species might arise from existing ones.[22] On the basis of a series of experiments on all the plants reported to be mutable, he reversed his earlier position on the production of new plant types and was even firmer than Duchesne in denying the reality of species mutability. Indeed, he accused Duchesne of believing in species mutability and camouflaging that belief with his use of the word "race." "The transmutation of species," Adanson concluded, "does not happen among plants, any more than among animals, and there is not even direct proof of it among the minerals, following the accepted principle that constancy is essential in the determination of a species."[23] He acknowledged that the characters of an individual might depart in some measure from the species type, but he assumed that "these deviations have also their laws and their limits: indeed, the more one observes, the more one convinces oneself that these monstrosities and variations have a certain latitude, necessary no doubt for the equilibrium of things, after which they return into the harmonic order pre-established by the wisdom of the Creator."[24] In this same period, the 1760s, the German botanist Joseph Kolreuter concluded from extensive experimental studies that hybridization was not capable of producing new species: rather than establishing new lines of their own, hybrids tended to revert to parental types.[25]

Duchesne and Adanson were indebted to Buffon for their ideas on species.[26] In the *Histoire naturelle,* specifically in the article on the ass (1753), Buffon maintained that what counted most in defining species was reproductive continuity. The species, he wrote, "is nothing other than a constant succession of similar individuals that reproduce themselves."[27] He raised the issue of common descent: "If it were admitted that in the animals, and even in the plants, there were, I do not say several species, but a single one that had been produced by the degeneration of another species; if it were true that the ass is only a degenerated horse; there would be no more limits to the power of Nature, and one would not be wrong in supposing that from a single being she was able to derive with time all the other organized beings."[28] But there were no cases, Buffon went on to insist, in which one species had clearly degenerated from another. It could not be concluded that nature had derived all the different living things from a single one. No new species had appeared since the time of Aristotle, Buffon maintained, and the species that did exist were separated from one another by barriers of sterility.

Buffon later relaxed his views on the limits of organic mutability. As he studied the geographical distribution of animals, and as he began thinking about an extended geological time scale, he came to acknowledge that organic forms could change appreciably. In an article on the animals common to the Old and New Worlds (1761), he maintained that there was something about the physical conditions of the New World that prevented life from developing as fully as it had in the Old World: the largest animals of the New World could not compare in size with the largest animals of the Old, the animals common to both were smaller in the New than in the Old, the domesticated animals that had been transported from the Old World to the New had decreased in size, and the natives of the New World displayed less vigour and virility than their European counterparts.[29] Later, discussing "the degeneration of animals" (1766), he identified three main causes of the changes in living things: climate, nutrition, and domestication.[30] He maintained that the races of man had all descended from the same stock, as evidenced by their interfertility, and that their differences were due to the different environmental conditions to which they had been subjected. If Negroes were transported to Denmark, he suggested, they would become white (again), though the change would not be rapid.[31] On a more general level, he proposed that the two hundred species of existing quadrupeds could be grouped into thirty-eight families representing thirty-eight original stocks from which the present day species of quadrupeds had descended.[32] He thus espoused quite explicitly a view of descent through modification. His view, however, presupposed neither the derivation of all forms from a single one nor the development of complex forms from more simple ones. What it presupposed was the existence of certain "internal molds" that were virtually part of the structure of the universe and that determined the basic types of life appearing on the earth.[33] The basic types of life appeared through a process of spontaneous generation, one group of life forms appearing in the third epoch of the earth's history, and a second group of life forms appearing in the fifth epoch of the earth's history, after the first group had been annihilated by the cooling of the earth's temperature.[34]

Original as Buffon's thinking was, it can be seen as representative of the understanding of organic change that was common to the naturalists of the eighteenth century. Unable to emancipate themselves from the idea of essential organic types, the naturalists assumed these types were either the direct result of the Creation (Linnaeus' view) or the fundamental molds of generation (Buffon's view). They supposed that if new forms did arise over time, these forms were either hybrids or degenerations of type. They did not assume that organic change was fundamentally progressive or that it ever proceeded beyond the bounds of certain fixed, primordial types.

The notions of progressive and unlimited change did find expression in the eighteenth century, though not in the writings of the naturalists. For example, the *philosophe* and physicist P.-L.-M. de Maupertuis, advancing a theory of generation, went so far as to suggest that "from two individuals alone the multiplication of the most dissimilar species could have followed."[35] Maupertuis failed to indicate whether the changes he had in mind were to be conceived as *progressive*. There is no mistaking the fact, however, that he toyed with the idea of an organic change that was virtually *unlimited*.

Maupertuis felt that only two basic hypotheses had been offered concerning the origin of living things: either living things had been formed by the chance association of elements or else they had been built from the elements by the "Supreme Being" (or His agents) the way an architect builds an edifice from stones. Maupertuis offered a third hypothesis, namely, that "the elements themselves, endowed with intelligence, arrange themselves and unite with one another to fulfill the views of the Creator."[36]

In his theory of generation, Maupertuis proposed that each part of the body furnishes particular germs to the semen of the parent, and that, in the formation of the fetus, germs from each parent unite to reproduce the part from which they were derived. With this idea Maupertuis was able to explain easily the phenomenon of biparental heredity, which had posed special difficulties for those who believed that the organism was already preformed in the egg or sperm. He was also able to explain monstrosities, another difficulty for the performationists. Monstrosities, he proposed, arose as the result of "chance combinations of the particles of the seminal fluids, or the effects of affinities between these particles that are too strong or too weak."[37] The naturalists of Maupertuis' day recognized that forms differing from the parental type sometimes arose suddenly. They supposed, however, that such variations were accidents, not only in the sense of being chance events but also in the sense of having nothing to do with the essential characters of the species. Maupertuis, in contrast, suggested that new species could be produced by chance if the hereditary particles did not retain the order they held in the parent organisms: "Each degree of error would have made a new species, and by dint of repeated deviations would have brought forth the infinite diversity of animals that we see today."[38]

Maupertuis' views did not go unnoticed. Diderot and Buffon in particular paid special attention to them. But Diderot, believing that a more genuinely materialistic explanation of the universe was possible, could not accept an explanation of generation that invested matter with the properties of desire, aversion, memory, and intelligence, as Maupertuis' theory seemed to do.[39] And Buffon, while impressed by Maupertuis' ideas, was prevented by his own views on generation from accepting the idea that all the different species could have descended from a single prototype.

Maupertuis wrote, if only briefly, about transformations of species. Not genuinely transformist, but endorsing the idea of organic *progress* in a way that Maupertuis did not, were the theories offered by the French philosopher J.-B. Robinet and the Genevan philosopher-naturalist Charles Bonnet. Robinet's thought is a prime example of what Arthur O. Lovejoy has called "the temporalizing of the Chain of Being."[40] Robinet believed in a unity of plan in nature's productions. "Each variation of the prototype," he wrote, "is a sort of study of the human form that nature is meditating upon."[41] Robinet supposed that the higher forms of life were produced only after the lower forms. He did not suppose, however, that a given species in the chain of being actually gave rise to the next species higher up in the chain. In his system, the filiation between successive forms was wholly ideal.

Bonnet offered a theory of progressive organic development in his *Palingénésie philosophique, ou idées sur l'état passé et sur l'état futur des êtres vivans* (1770).[42] In earlier writings Bonnet advanced a belief that the species had been fixed since the Creation, but his theory of palingenesis allowed for organic change over time. All organized beings, he supposed, contained germs that were unsuited for development in the present stage of the world but which would develop after the world underwent further revolutions and the conditions of existence changed. This development was foreordained: the germs were originally created to correspond to the diverse coming revolutions.[43] Bonnet supposed that when the revolutions of the globe called forth the development of different germs, the resultant organic changes were essentially progressive. These changes did not involve one generation giving birth to another, however. They were reincarnations. New conditions of the earth's surface allowed individuals to rise again in improved form from germs contained in a dormant state in their former bodies. Bonnet was as much concerned with resurrection and the nature of the soul as he was with organic development. Although Bonnet's system was not one of biological evolution in a modern sense, it was, nonetheless, a system in which organic change took place over time, and the change was essentially progressive.

IN THE WORDS of Charles Darwin, "the belief that species were immutable productions was almost unavoidable as long as the history of the world was thought to be of short duration."[44] At the beginning of the eighteenth century, the earth was commonly estimated to be only six thousand years old. This estimate was based on the account of the earth's history contained in the Bible. By the middle of the century, however, increasing attention to fossil evidence had begun to cast considerable doubt on the scientific validity of the Scriptures in geological matters. Prominent among those who contemplated the meaning of fossils were the leading naturalists at the Jardin du Roi: Buffon, Guillaume-François Rouelle (best known in the history of science as the teacher of the great French chemists of the latter part of the nineteenth century), and the botanist Bernard de Jussieu. They all entertained the idea that the earth was immensely older than had generally been supposed.[45] Also attentive to the evidence of fossils was the notorious de Maillet, who scandalized the eighteenth century with his idea that the human race had come from mermen and mermaids, but who provided a number of shrewd observations on the earth's history.[46]

Fossilized remains of marine animals had been discovered in massive beds far distant from any modern seas. Some eighteenth-century writers explained the location of these fossils by the Deluge, but de Maillet and the naturalists pointed out that the condition of the fossil shells indicated that the shell beds had been laid down gradually over long periods of time, not tossed into one place by a catastrophe such as Noah's flood. Buffon, Rouelle, and Jussieu also found special significance in the fact that the fossil fauna and flora of their region was by no means identical to the modern fauna and flora of the same locality. They noted in particular that when living analogs of the marine fossils of France were found, these were typically discovered in distant, tropical waters. The naturalists concluded that not only had the oceans shifted their location in the course of the earth's history, but specific areas of the earth's surface had been subjected to major climatic changes. As Jussieu put it: "a new world had come to form itself upon the old."[47]

The building up of vast beds of shells or coral could be explained by everyday processes operating over long periods of time. The displacement of the ocean beds and climatic changes on the earth's surface could be accounted for in similar terms. Buffon offered a causal mechanism for the gradual displacement of the sea beds: the combined effect of the gravitational pull of the moon and the rotation of the earth on the movement of the earth's waters (Lamarck adopted the fundamentals of this explanation, without acknowledgment, in his *Hydrogéologie* of 1802). According to Buffon, the ocean beds are sculpted and maintained by the oscillatory motion of the tides. As the earth rotates on its axis from west to east, the ocean waters, which are subject to the same general motion, are retarded in their movement by the attractive powers of the moon. The result is a small and slow, but nonetheless definite, motion of these waters toward the west. Gradually, as the result of the erosive action of the waters, the east coasts of continents are invaded while the west coasts are correspondingly abandoned, and the beds of the oceans are very slowly displaced.[48]

Climatic changes on the earth's surface could be explained by changes in the earth's ecliptic. Rouelle speculated on the great changes that had taken place on the earth's surface and concluded that "only by the action of a powerful cause that has produced and produces these changes each day, slowly," could one account for the "revolution" in the earth's past.[49] The cause Rouelle had in mind was the gradual displacement of the earth's axis, from which the change of climates and displacement of the seas would follow. The Baron d'Holbach, a frequenter of Rouelle's lectures at the Jardin, echoed Rouelle's views on how changes in the location of the seas and climate could result from the displacement of the earth's axis:

> For such a large part of a continent [once under water] to become dry land, a very considerable revolution was necessary. According to the most probable opinion, this came from the nutation of the earth's axis and the change of the inclination of the ecliptic on the equator, occasioned by the change of the earth's center of gravity. These events, recognized by the majority of physicists, have been sufficient to produce the most marked

alteration at the surface of our globe: they must have not only caused the waters of the sea to disappear from the places where they were in order to go submerge others, but they must also have altered the whole position of the globe relative to the sun, consequently causing a total change in the climate and influencing the individuals found there.[50]

De Maillet, Buffon, Rouelle, Bernard de Jussieu, and d'Holbach were not alone in their belief in the earth's great antiquity, which they based on fossil evidence. There was much more speculation on the earth's history than found its way into print. One enticing hint of such speculation comes from a geological memoir written by Lavoisier. In regard to what he called the "strange finding" that the chalk is ordinarily the last of the strata that contains animal remains, he wrote:

If it were permitted to hazard some conjectures on this strange result, I would believe it possible to conclude from it, as M. Monge was the first to do, that the earth has not always been populated with living things; that it was for a long time an inanimate desert where nothing had life; that the existence of plants preceded by a long time the existence of animals, or at least that the earth was covered with trees and plants before the seas were populated with shells. In a sequel [to the present memoir] I will discuss in very great detail these opinions which belong much more to M. Monge than to myself.[51]

Unfortunately, the sequel to Lavoisier's first geological memoir never appeared, and Monge's ideas on the history of life on the earth were never recorded in detail.

Other views on the history of life that remained unpublished in the eighteenth century and have presumably been lost forever include those of the Abbé Soulavie. In the first volume of his *Histoire naturelle de la France méridionale,* published in 1780, Soulavie instructed his readers to consult in the following volumes "the natural history of the Mediterranean, where I will prove that the recent families [of shellfish] that are not found in the old world marbles are nevertheless species that have descended from primordial families."[52] The natural history of the Mediterranean did appear, but the discussion of descent with modification did not. Soulavie's views on the latter subject were censored.

It is not known what sort of theory of organic change Soulavie would have presented had he been allowed to do so, though his mention of "primordial families" suggests a framework similar to

Buffon's. It is plausible at least that had Soulavie presented a theory of organic change, he would have drawn heavily upon his own experience in the field as a geologist and paleontologist. He made the important observation that different strata are characterized by different fossils and also noted that fossil forms become increasingly similar to modern forms as one moves from the lower to the upper strata of the earth's crust. The prospectus of his work included a chapter entitled: "Views on the ancient history of the organized world. Comparison of the monuments of this age. Metamorphoses of several species of animals. Comparison of the fossil shells of the first three ages."[53] The chapter never appeared.

Though the importance of an extended geological time scale for the development of evolutionary biology cannot be overestimated, in the eighteenth century simply granting the earth a great age was not sufficient to give life a fundamental history of its own. According to Foucault, it was impossible for natural history to conceive of the history of nature: in the systems of Robinet and Maupertuis "succession and history are for nature merely means of traversing the infinite fabric of variations of which it is capable."[54] Lamarck, despite his work as a systematist which led him to compare fossil and modern shells, failed to think of the history of life in the way that most biologists of the nineteenth century would think of it. He assumed that the order nature had followed in bringing the different animals into existence was essentially the same as the scale of increasing complexity he used to classify them. In addition, his evolutionary theorizing laid special emphasis on how organisms would develop *naturally* were it not for the constraining accidents of history. He never attempted anything like Soulavie's apparent project of tracing changes in fossil forms across successive geological strata. The most important stratigraphical research done in France at the beginning of the nineteenth century was conceived and carried out by Cuvier and Alexandre Brongniart, not Lamarck.

LAMARCK'S EARLY BELIEF IN THE IMMUTABILITY OF SPECIES

Lamarck's botanical writings of the eighteenth century bear witness to the validity of his assertion in 1802 that he had "thought for a long time there were constant species in nature."[55] In his *Flore françoise,* in his three and one-half volume contribution to the botanical part of the *Encyclopédie méthodique,* and in his various botanical articles he never challenged the idea of the immutability of species. His early thoughts on the "natural" way to arrange nature's productions eventually proved central to his conception of the evolutionary process. Nonetheless, the order of nature as he initially conceived of it did not require a dynamic interpretation. Furthermore, his characterization of species and the keys he constructed to facilitate their identification fit nicely with the view of species as fixed types.

In the *Flore françoise* and in his other botanical writings prior to 1800, Lamarck insisted upon the reality of species in nature. Though he regarded the higher categories of classification as merely the constructs of human intellect, the species, he believed, genuinely existed. Constant characters distinguished them from one another.[56] He was well aware that individuals of the same species could differ appreciably from one another, causing the botanist major difficulties in determining whether a new form was an unknown species or simply an unknown variety. But such difficulties, Lamarck maintained, did not allow one to conclude that the species, like the higher categories, existed only in the mind of the botanist and not in nature. The duty of the botanist was to avoid becoming lost in individual differences and to focus instead on the essential characters of the species: "The individual that offers itself to the botanist in his researches is not in his eyes an isolated being; he sees there the type and the model of the entire species."[57] Whatever difficulties the botanist might have distinguishing species and arranging them in a systematic fashion, he could sustain the belief that "models [of the species], among those of all the possible creatures," were "arranged without confusion in an infinite intelligence."[58]

Following Buffon, Lamarck maintained that the surest criteria for determining whether or not species status should be accorded to any particular form were constancy through reproduction and infertility with other forms. In the *Flore françoise* he complained that the majority of his contemporaries were bestowing species status upon mere varieties:

> Instead of trying to distinguish the species by well-defined characters, always confirmed by constancy in reproduction, and without ever recognizing mere differences of degree [*sans jamais employer le plus ou le moins*], nearly all present-day botanists are multiplying the species, at the expense of their varieties, to infinity. The slightest nuance in the size, the color, or the consistency of two individuals is all they need to form two different species. They pay no attention to the fact that the seeds of a single plant,

> if taken to two different places and exposed to and cultivated under entirely opposite circumstances, will necessarily produce, after a few years, two plants that will differ greatly in their external appearances.[59]

In the definition of species that he provided for the *Encyclopédie méthodique,* Lamarck stressed the importance of reproductive continuity and the distinction between essential and accidental characteristics:

> In Botany as in Zoology, the species is necessarily constituted by the whole group of similar individuals that perpetuate their kind through reproduction. By similar I mean in the qualities essential to the *species,* because the individuals that belong to it often display accidental differences that are the basis of varieties, and sometimes display sexual differences.[60]

The differences in appearance between two individuals, Lamarck insisted, were not always trustworthy indices of whether the two individuals represented the same or different species: "In the determination of species, one must often pay less attention to the size of the differences presented by the individuals under examination than to the constant conservation of these differences in reproduction by seed. I am convinced of the fact that two species constantly distinct in reproduction sometimes offer less differences between them than do two varieties of the same species."[61]

It was by no means an easy matter, Lamarck admitted, to know "the true characters of the species." Ideally, he thought, the botanist should observe plants only in their natural habitats—not in gardens "where they are often altered by borrowed traits."[62] It was impossible to know a plant completely unless one saw it "in the state that is natural to it."[63] That, Lamarck pointed out, was why herborizations were so important. He did not suppose that botanical gardens were entirely useless to the botanist, though, because he assumed domestication or cultivation could not change the essential or natural characters of a plant.[64]

In a discussion of wheat (1788) Lamarck indicated just how far he believed the individuals of a species could be modified by the influence of the environment.[65] It was no surprise, he said, that there were so many varieties of wheat. The wheat of Egypt, Italy, and Spain grew under a different climate than the wheat of Denmark and Sweden, and not only did wheat exhibit variations according to the different climates in which it was grown, it also exhibited variations resulting from its subjection to different soils, different exposures, and different methods of cultivation. Linnaeus regarded winter and spring wheat as two different species. Lamarck supposed Linnaeus was mistaken. "It is not doubtful," Lamarck claimed, "that by sowing spring wheat in September for a certain number of years it would be turned into winter wheat and by sowing winter wheat in March for a certain time, it would be turned into spring wheat."[66] Lamarck was prepared to admit that the individuals of a species, in forming varieties, could depart considerably from the basic species type. He assumed, however, that circumstances could not effect change beyond the limits of a species. The species themselves remained unchanged.

Many of the ideas that Lamarck later employed in his explanation of the mechanisms of evolution had been previously accepted in a nonevolutionary context by himself and others. Such ideas are to be found, for example, in an article of 1777 by the Baron de Tschoudi, whose writings Lamarck admired.[67] Discussing the subject of "acclimatization" at some length in the supplement to the *Encyclopédie,* de Tschoudi made a number of comments reminiscent of views Lamarck employed in setting forth his evolutionary theory in the 1800s. Analyzing the effects of climate and soil upon transplants, de Tschoudi observed that a plant's *habitude* had to be altered gradually if transplantation were to be successful. Organic changes occur, he said, as the result of factors that operate gradually and continually upon supple organs. He did not doubt that such acquired changes could be passed down to succeeding generations. He discussed the production of new varieties through hybridization. He also recommended the establishment of special societies to study "the reproduction, the transformations, and the processes of perfection of plant life."[68] But de Tschoudi advanced no theory of evolution in his article. Organic forms could be perfected without being changed in their essentials.

Questions about the constancy of species continued to be raised in the 1790s. In 1792, in the *Journal d'histoire naturelle,* of which Lamarck was a co-editor, Louis Reynier, whom Lamarck had proposed for membership in the Société d'Histoire Naturelle early in the same year, wrote a long article entitled "on the influence of climate on the form and nature of plants."[69] Among other examples, he cited how *Ranunculus aquatilus* exhibits dramatically different forms according to the conditions under which it grows. In the same journal, Antoine-Nicolas Duchesne, in another discussion of his Versailles strawberry, avowed that he could not yet see that "the spiny question of species constancy" was close to being resolved.[70] He went so far as to ask whether the "privilege of immutability" might have to be reserved for the genera rather than the species. But Duchesne does not appear to have been thinking in terms of anything beyond a very limited sort of organic mutability, and the same may well hold true for Reynier, despite the fact that Reynier was an outspoken opponent of the doctrine of the *emboîtement* of germs and the "Linneists' " supposition that all plants come from seeds.[71] Reynier believed in the spontaneous generation of the simplest forms of life. It is not apparent, however, that he believed in the transmutation of species. His views appear in many respects to resemble those of the editor of the *Journal de physique,* Jean-Claude Delamétherie, who supposed that the different forms of life arose initially through crystallization but who denied that organic mutability could proceed beyond the limits of the species. The importance of Reynier's work on the influence of climate on plant form, as Delamétherie himself signaled it, was not that it demonstrated species change. To the contrary, it seemed to Delamétherie that given the "considerable changes" in plants Reynier had documented, the botanist had to be careful not to take for a new species what was only a variety produced by the soil or climate.[72]

After 1800 Lamarck used Reynier's example of *Ranunculus aquitilus* as a prime case of the role of the environment in the production of new organic forms. In 1792, however, Lamarck was content with the conventional view that species remain constant despite the changes particular individuals of the species undergo. His position at that point may be summarized by the following propositions:

1. Species really do exist in nature, unlike the higher categories of classification, which are the result of arbitrary decisions on the part of the naturalist.
2. Essential differences separate the species from one another.
3. The truly solid part of natural history, consequently, is a knowledge of the species, and the description of new species will provide the principal base for the progress of natural history in the future.[73]

In his years as a botanist Lamarck picked up ideas concerning the nature and causes of individual variation, and he eventually used these notions in his explanation of evolution. But inasmuch as he and many others were able for many years to embrace these ideas within a view of species as fixed types, it appears that the ideas in themselves were probably insufficient to convert Lamarck to a broad-scale understanding of organic change. He and his contemporaries could speak of the question of species constancy as being "spiny." He could believe that the environment induced changes in living things, and that acquired characters were inherited. Yet he was still far from conceiving his general theory of evolution. It is inappropriate to regard Lamarck's thoughts on organic change as a simple extension of the ideas of Duchesne, Reynier, or Buffon. Lamarck's thoughts were set in a different framework.

IN REVIEWING the major eighteenth-century ideas about organic mutability and the origin of species, it is evident that there were important exceptions to Lamarck's statement in the *Philosophie zoologique* that nearly everyone before him believed in species fixity. One could dismiss, of course, a fanciful thinker like de Maillet, who proposed that flying fish could change directly into birds.[74] One could also dismiss, no doubt, a Maupertuis. How was one to believe that so many different forms, so well adapted to the conditions of their existence, could ever have arisen through a series of accidents? But it could not go unnoticed that the two most prominent naturalists of the eighteenth century, Linnaeus and Buffon, both set forth explanations of organic diversity involving the production of new species over time. Linnaeus proposed that God had brought certain living things into existence at the Creation and that from these, through hybridization, numerous other living things were produced. Buffon proposed that a certain number of primordial forms arose by spontaneous generation when the earth's surface became fit for life and that the "degenerations" of those forms that had been spontaneously generated in the fifth epoch of the earth's history had resulted in all the different forms alive in the eighteenth century. That the Linnean and Buffonian models of organic change were not forgotten by the end of the century is evidenced by an interesting statement offered in a joint article by the young naturalists Georges Cuvier and Etienne Geoffroy Saint-Hilaire in 1795:

It is a rather general law of nature that the number of species in a genus is approximately proportionate to the fecundity of each of them; *whether what we call species are only the diverse degenerations of the same type, which have had to multiply themselves in proportion to more or less frequent birth; or whether many of them are born from the coupling of neighboring species, the efficiency of this sort of mixing depending on the generative force of the species mixed.* (Italics added)[75]

Lamarck, in the evolutionary theory that he developed at the beginning of the nineteenth century, stressed the importance of environmental change as a stimulus for organic change or degeneration. He also mentioned hybridization as a means by which new races might arise, though this mechanism never played a central role in his theory. In marked contrast to either the degeneration model of Buffon or the hybridization model of Linnaeus, however, the explanation of organic diversity Lamarck offered involved full-scale organic mutability. Rather than supposing that an assortment of primordial forms of different complexity had hybridized or degenerated to produce all the other forms, he supposed that nature had begun with the very simplest forms and from these had successively produced the rest. That Lamarck's theory took the form it did was dependent both on the intellectual habits he had developed in his early years as a naturalist and on the special field of study that became his concern and responsibility in the 1790s—the zoology of the invertebrates.

CHAPTER FOUR

The Preoccupations of the New Professor: Chemistry, Meteorology, and Geology

Though certain fundamental assumptions of Lamarck's early thought contributed significantly to the evolutionary theory he eventually developed, and though various facts and ideas available to him in the first part of his career found expression when he later wrote about organic change, there is nothing in his early writings to suggest that had he remained primarily a botanist—even one with a broad interest in the fundamental processes of nature—he would ever have come to believe in evolution. For the inspiration and development of his evolutionary views, the new concerns deriving from his appointment in 1793 as professor of "insects, worms, and microscopic animals" at the Muséum d'Histoire Naturelle were of central importance. But his conversion to a belief in evolution did not follow immediately upon his assumption of the new professorship. Not until 1800 did he endorse a broad, evolutionary explanation of how the different forms of life originated.

In order to understand the evolutionary theory Lamarck eventually developed, it is important to know what inspired his evolutionary thinking in the first place. Identifying the inspiration of Lamarck's views, however, is no simple matter. Lamarck himself did little in the way of reconstructing how his ideas originated. In addition, precisely when he adopted his views on organic change is not certain. His last explicit endorsement of species fixity occurred in 1794. His first explicit endorsement of species mutability came in 1800.[1] To complicate matters further, his interests in the period 1794–1800 were extremely diverse. He tackled problems in chemistry, meteorology, and geology, as well as in zoology, and maintained that these subjects were all related to one other.

The difficulties posed in the 1790s by the classification of the "insects, worms, and microscopic animals" would have been enough to keep most naturalists busy. This part of the animal kingdom had received the least attention from systematists throughout most of the eighteenth century, and at the beginning of the 1790s, the field remained in a state of general confusion. But Lamarck, as just suggested, did not devote himself single-mindedly to his new charge in 1794. He also pondered the causes of many of the basic phenomena of chemistry, meteorology, and geology and contemplated how to establish these sciences on firm foundations. Indeed, judging from his publication record in the half-dozen years prior to his first enunciation of his evolutionary views, it appears that for much of the late 1790s, Lamarck was preoccupied with nonbiological subjects. Augustin Pyramus de Candolle recalled that as a young student seeking botanical instruction in 1796 he established contact with Lamarck but was not satisfied with the result: "I was not long in perceiving that my work with Lamarck was of little use to me. This scientist was then absorbed in his writings against modern chemistry and by his hypotheses relative to the action of the moon on the atmosphere. When I questioned him on botany, he responded with chemistry or meteorology, which he scarcely knew. I therefore gradually ceased going to see him."[2]

Following the negative results of his attempt to gain a hearing for his physico-chemical views at the Institut de France, Lamarck turned his attention more and more to the study of natural history. The papers he had initially intended to collect under the title "Memoirs presenting the bases of a new theory, physical and chemical" were collected instead under the title "Memoirs of physics and natural history."[3] In presenting the completed collection to the Société Philomathique, in 1797, he described it as a book treating physics and natural history "nearly equally" and serving "to make more apparent the connection that exists between these two sciences."[4] Actually, of the seven memoirs in the book, only the last one involves an extended discussion of plants, animals, and minerals. Lamarck evidently felt that the successful promotion of his book depended upon his playing up the part of his work devoted to natural history at the expense of the part devoted to physics and chemistry. If he was also falling back on his studies as a naturalist, however, this was only a partial retreat and it was made grudgingly. He did not renounce his physicochemical views. He also retained high hopes for the work in meteorology that he had begun years earlier but had not had the opportunity to develop to his satisfaction. Seventeen ninety-eight saw his first publication on meteorology, a memoir "on the influence of the moon on the terrestrial atmosphere."[5] In 1799 he published memoirs on "the substance of fire" and "the substance of sound," he offered to the public the first of his meteorological annuals, in which he presented "probabilities" regarding what the weather would be like throughout the coming year, and he delivered to the first class of the institute a memoir on fossils and their implications for geological theory.[6] In the same year he published two memoirs on zoological subjects, one describing the features of the cuttlefish, squid, and octupus, and the other setting forth a new classification of shells.[7]

It appears, in short, that when Lamarck first announced his evolutionary views in 1800, he was devoting more time to problems related to invertebrate zoology than he had in the previous years, but he was displaying considerable interest in geology and meteorology as well, and he had by no means forgotten the physico-chemical issues that caused him so much difficulty with his contemporaries. The nature of Lamarck's various interests in the 1790s and their relation to the biological and evolutionary views he developed in the 1800s remain to be explored. The present chapter will be devoted to a consideration of his physico-chemical, meteorological, and geological thought and its possible bearing on his biological theorizing. The following chapter will be devoted to his studies as an invertebrate zoologist and to what immediately inspired his views on organic evolution.

LAMARCK'S PHYSICO-CHEMICAL SYSTEM

Lamarck described his *Recherches sur les causes des principaux faits physiques* of 1794 as a "logique physico-chimique" and introduced it saying "I come supported by all the known facts and published experiments, which now belong to me as much as to their original authors. I come . . . to examine the proper consequences of these facts, and to propose my opinion on their results."[8] The full title evidenced the work's immense scope:

> Researches on the causes of the principal physical facts; and particularly on those of combustion, of the elevation of water in a vaporous state, of the heat produced by solids rubbing together, of the heat perceptible in sudden decompositions, in effervescences, and in the bodies of many animals during the course of their lives; of the color of bodies, of the origin of compounds and all the minerals, finally, of the maintenance of the life of organic beings, their growth, their state of vigor, their decline, and their death.

Lamarck recognized that the enterprise he had undertaken in his *Recherches* was vast, but, he pointed out, he had actually exercised some wise restraint in defining his subject. He had not sought "to explain the formation of the universe, to go back directly to the first causes of all that is, or to aspire to determine the true laws that have given rise to the existence of the beings we observe, or which have occasioned the reunion of the substances that compose them."[9] He refrained from treating these subjects because he supposed they were inaccessible to the human intellect. It seemed to him that there were three principal things that "man reasoning philosophically" would never know about: first, the cause that produced matter with all its essential qualities and faculties; second, the cause of "the existence of organic beings and that which constitutes the life and essence of these beings, since matter with all its qualities seems . . . to be in no way capable of producing a single being of this nature"; and third, the cause of "the *activity* that is spread throughout the universe."[10] Later in his career he would still include matter and its properties, the universe, and nature and nature's laws among the "observable objects whose origin is necessarily unknown."[11] The existence of organic beings and the nature of life, on the other hand, would no longer appear incomprehensible to him.

Four general considerations formed the foundation of Lamarck's *Recherches* of 1794 and continued to be the major themes of his physico-chemical thought for the rest of the decade.[12] The first related to fire (which Lamarck, upholding the four-element theory of traditional chemistry, considered to be one of the basic elements) and to its different states: a natural state, a "fixed" state, and a state of expansion. Fire, as Lamarck put it in 1797, had been a "reef" on which the quests of physicists and chemists had often been stranded.[13] He believed he had perceived how fire, in its natural state and its two states of modification, was central to a host of phenomena: not only combustion, calcination, evaporation, and animal heat, but also sound, color, causticity, and perhaps even electricity and magnetism.[14] Appropriately, he chose to name his theory the "pyrotic theory," in contrast to the "pneumatic theory" that the chemists of his day favored.[15] It is not necessary here to elaborate upon Lamarck's explanation of the various ways in which "the matter of fire" functioned in its three principal states or got from one state to another.[16] But it is important to note that when Lamarck sought a mechanical explanation of the origin of life, he found it readily in "the matter of fire"—the agent that was central to his physico-chemical system.

The second major theme of Lamarck's physico-chemical system was that all compounds have a tendency to decompose. Lamarck observed this tendency in the decaying remains of dead organisms and concluded that it was universal.[17] The elements, he was confident, had "no aptitude to modify themselves to form compounds."[18] For each of the four elements or elementary "principles," the state of chemical combination was a state of "constraint," "a state that in some measure deteriorates the very nature of the principle, in modifying it and enchaining its faculties."[19] It was unreasonable, Lamarck maintained, to suppose that any element would have a *tendency* to place itself in such a situation of constraint, to remove itself from its natural state and lose some of its faculties. Thus, some special cause was necessary to form compounds, since the natural tendency of elements was to disengage themselves from the state of combination whenever they had been forced into it. This was the kind of "base of reasoning, independent of all theory" on which he felt chemistry ought to be founded, the kind of base that would serve "to put a check on the excessive inclination of certain men to build systems."[20]

The third concept central to Lamarck's physico-chemical thinking was his notion of what he called the "essential molecule" of any compound. His definition of the essential molecule was unexceptional enough: "the smallest molecule to which the mass of a compound can be reduced, without the nature of this substance being altered," resulting from "the combination of a certain number of principles united together in certain proportions."[21] He broke with contemporary chemical theory in his view of what happened to this essential molecule in chemical reactions. Assuming that the tendency of all compounds was to decompose, he denied the contemporary notion of chemical *affinities* and claimed that in all chemical operations in which compounds were combined or one compound was derived from another the original substances were denatured.

The final major element of Lamarck's physico-chemical system was his identification of "the principal productive cause of all the compounds that exist." This cause, he explained, "resides in the organic action of living beings, and particularly in the most surprising of their faculties: *nutrition*."[22] In living things, he declared in his *Recherches* of 1794, there are "two powerful and very distinct forces, always in opposition, and mutually combating each other without cease, in a manner that each of them destroys perpetually the effects that the other manages to produce."[23] One of these was a tendency he attributed to nature: the tendency of all compounds to decompose. The other was a tendency he attributed to what he called the "power" or "principle" of life. In 1794 he considered the principle of life to be apart from nature because "its origin and essence cannot be assigned physically."[24] Only this principle of life could form compounds. Were it not for life, there would be no minerals or any other compounds, since the different minerals were formed as the remains of organic beings decomposed. Lamarck acknowledged, in the *Recherches,* that this posed certain problems regarding the origin of the world: namely, how the first living things were able to exist, and whether there were minerals at the world's beginning. He dismissed this problem simply be indicating that it was not his role to speculate about what could not be verified: "I have limited myself in these researches to the observation of nature such as it is presently."[25]

A brief look at the foundations of Lamarck's physico-chemical system indicates that biology played a key role. Lamarck indicated on numerous occasions that he considered his physico-chemical and his biological views to be closely related.[26] The question that inevitably confronts the historian of evolutionary thought is what sort of role, if any, Lamarck's physico-chemical system had in the development of his evolutionary thought. None of the four major themes of Lamarck's physics and chemistry leads necessarily to the idea of evolution in biology. To the contrary, for Lamarck to develop his broad theory of organic development he had to give up one of the central assumptions of his physico-chemical system: his assumption that life could never be produced from non-life.

The importance of Lamarck's physico-chemical speculations for the development of his evolutionary views was not, as claimed by one historian of science, that "the first statement of what became Lamarck's evolutionary theory occurs in his assertion of the indefinite variability of chemical composition," or that "what Lamarck did between 1797 and 1800 was to assimilate the question of organic species (or rather of their non-existence) to that of species in general, and of mineral species in particular."[27] A concept of "species in general" was never important to Lamarck, and he insisted throughout his career (as did his colleague Daubenton) on the essential distinction between the "species" of the organic realm on the one hand and the "kinds" [*sortes*] of minerals on the other. Lamarck occasionally used the word "species" for minerals as well as living things, but he remained keenly aware that "the species among inorganic bodies must be distinguished from that of living bodies, because in the two cases the definition of species and the source of the individuals is very different."[28] What was important in Lamarck's physico-chemical views of the 1790s, for the evolutionary theory he began expounding in 1800, was the development of his thought on the nature of life, certain patterns of explanation that were transferred consciously or unconsciously to his treatment of the production of organic diversity, and his specific concern with the presence and activity of fire in its "state of expansion" (the "caloric" of many of his contemporaries).

Of all the subjects he treated in his physico-chemical works of the 1790s, Lamarck felt that those relating to the life processes were the most interesting.[29] Between 1794 and 1797 his statements on the nature of life changed in an important way. In both 1794 and 1797 he defined life as organic movement, but while in 1794 he maintained that "that which constitutes the essence of life in an organic being is truly a principle forever inconceivable to man,"[30] in 1797 he was no longer speaking of life as an inconceivable *principle*. Instead, he stated explicitly that he did not consider the essence of life in a body to reside in a particular being, or soul, which vivified the body. The religious issue was one which he, as a scientist, felt perfectly justified in skirting. Neither "man's immortal soul" nor "the perishable soul of beasts" could be known physically.[31] Life, on the other hand, could be known physically. As he defined it in 1797: "Life . . . is nothing other than the movement in the parts of [organized] beings resulting from the execution of the functions of their essential organs, or the possibility of being in possession of this movement when it is suspended."[32]

It is not certain what prompted Lamarck to cease referring to life as a vital principle, or for that matter, how much his thinking on the nature of life changed during the course of his career. Prior to 1794 Lamarck had described life in plants in basically mechanistic terms, and as of 1797 he was using these same terms again. At any rate, phenomena displayed by some of the more simple invertebrates seem to have influenced Lamarck's thinking on life in the 1790s. Especially striking in Lamarck's new definition of life in 1797 was his insistence that life was defined by organic movement *or the possibility of exercising organic movement.* Lamarck's expansion of the definition to include the *possibility* of exercising organic movement may well have been inspired by Spallanzani's studies on the reanimation of dessicated rotifers. This is suggested by a comment Lamarck published four years later: "It has been observed that [rotifers], dried promptly, and consequently without any movement whatever and without active life, being conserved in this state for entire years even, but sheltered from any deterioration, can then, if put back in water, take up movement and life again. Spallanzani's rotifier (*Urceolaria rediviva*) served first to make this faculty known. How this singular fact enlarges our ideas, and what light it sheds on what is called *life* in all the beings endowed with it!"[33] Life in an organism could be suspended, but the organism was not dead until the *order of things* necessary for its life functions had irrevocably deteriorated.

Defining life as organic movement or the possibility thereof, Lamarck in 1797 distinguished between the causes of organic movement in plants and animals in much the same terms he had used the previous decade when discussing plant "sensitivity." Organic movement in animals, he said, was caused by an internal stimulus, the irritability of animal fiber. The faculty of irritability, he explained, "is the immediate cause of the reaction of the solid or containing parts on the essential fluids of animals; a continuous reaction provoked by the presence, the state, and the action of the fluids, from which arise the organic movements that constitute the life of these beings."[34] Organic movement in plants, on the other hand, was caused by an external stimulus. Plant fiber, Lamarck thought, was not irritable. Therefore, the fluids of plants could not stimulate the solid parts of plants. Lamarck analogized:

But if the irritability of the heart and arteries can communicate movements to the essential fluids of animals, causing them to circulate; the alternating variations of temperature of the surrounding air, caused by the constant succession of days and nights, can bring about particular movements in the fluids of these plants, alternating toward one end and then the other, and finally successive exhalations and absorptions, which constitute the vital movements of these beings.[35]

In 1797 Lamarck thus had a basically mechanical explanation for the organic motions of plants and a similar explanation, aided by the special faculty of irritability, for animals. He had also begun to think in terms of the effects of moving fluids on the development of organic form, explaining the formation of membranous tubes in plant parenchyma in this fashion.[36] The creative action of fluids in action was not prominent in his writings of 1797, but five years later, in his first major exposition of his evolutionary views, it was central. And insofar as Lamarck viewed subtle fluids to be responsible for setting the ponderable fluids of organic bodies into motion, the evolutionary theory he developed had a direct tie with his physico-chemical thinking: it called upon "the matter of fire" to serve in one form or another as the immediate cause of organic motion and hence of organic change.

The assumption that the motion of subtle and ponderable fluids alike was a key to understanding vital phenomena was not the only feature linking Lamarck's physico-chemical thinking of the 1790s with his ideas on organic evolution of the 1800s. They also had in common two broad patterns of thought. The first, already noted in Chapter Two, involves the model Lamarck offered to account for the origin of the minerals. His supposition that all the different minerals were produced gradually as the elements disengaged themselves from the remains of living things appears to be virtually the inverse of his later idea that all the different organic species were produced gradually as the "power of life" and modifying circumstances caused simple, spontaneously generated forms to become increasingly complex and diversified. The second pattern of thought common to Lamarck's pre-1800 and post-1800 periods involves the model he used initially to account for the phenomena of individual growth and development. He explained growth, maturity, and old age in the individual organism as the result of the interaction of the constructive power of life and the destructive power of nature. This found an echo in his later explanation of organic diversity as the result of the power of life, or tendency to increased complexity, on the one hand, and the constraining influence of circumstances (particular environmental influences), on the other.

Although there are notable continuities between Lamarck's physico-chemical views and his biological views, the fact remains that Lamarck's physico-chemical theories could never have led directly to his biological theories. Lamarck's chemistry was so exceptional in the last years of the eighteenth century because of his rejection of the common chemical notion of affinity of composition and because of his assumption that the tendency of all compounds is to decompose. On the basis of this theoretical framework, he could never have supposed that it was possible for life to arise spontaneously from unorganized, nonliving matter. Yet the idea that life is spontaneously or directly generated from non-life was essential to the development of his evolutionary theory. Clearly, Lamarck's biological views of the nineteenth century were not independent of the thoughts on the nature of life that he held in the eighteenth century. But the reasons he came to believe in spontaneous generation—which directly contradicted one of the fundamental bases of his physico-chemical system—and in the mutability of organic species cannot be explained simply by an examination of his physico-chemical ideas.

THE RELATIONS BETWEEN LAMARCK'S METEOROLOGY AND BIOLOGY

As indicated in Chapter Two, eighteenth-century theorists were quite prepared to acknowledge that atmospheric conditions could have a decided influence upon vital phenomena. The timing of Lamarck's increased attention to meteorological studies and the first announcement of his evolutionary views is sufficient reason to inquire whether his meteorological studies had any direct bearing on the development of his biological thought. Lamarck listed the influence of atmospheric phenomena on plants and animals as one of the three most essential kinds of meteorological knowledge, the other two being "the knowledge of the order that exists, however obscure it may be, in the succession of the large variations of the atmosphere in our climates" and "the circumstances that are essential to the formation of each atmospheric phenomenon."[37]

Lamarck in his earliest writings indicated that the phenomenon of life in plants depended upon the subtle fluids of the atmosphere. It was a major step in the development of his thinking when he assumed that the same held true for the simplest animals. But this insight does not seem to owe as much to his thoughts on atmospheric phenomena as it does to his reflections about the nature of life in the simplest animals. The main goal of his meteorological studies was to find and explain broad regularities underlying the day-to-day vagaries of the weather. He also attempted to account for particular meteorological phenomena: violent storms, cloud formations, unseasonable temperatures, and unusual dry spells. But he basically took for granted the influence of the atmosphere on living things and did not study it further.[38]

Lamarck's work in meteorology paralleled the rest of his work in a number of significant ways. Indeed, his intellectual behavior tended to be consistent whether he was studying meterology, chemistry, geology, or biology. Whatever the science in question, he believed it was up to him to establish it on a secure foundation, he was confident that a regular pattern underlay complex phenomena, and he exhibited feelings of paranoia when his efforts were not commended.

Lamarck considered himself to be the first scientist ever to set forth "a general system of meteorology," the first investigator to overcome the difficulties that had prevented the recognition of both the general and the particular causes responsible for changes in weather during the course of each year.[39] Moreover, he believed that his system of meteorology rested on a firm base and would never be overturned. His efforts, he supposed, marked both the beginning of meteorology as a science and the source of its progress in the future.[40] "The difficulties I have conquered," he immodestly announced, "are greater than those that remain to be overcome."[41]

The problem in studying atmospheric phenomena, as Lamarck perceived it, was that local "circumstances" obscured all too often the "natural" state of the atmosphere.[42] Here was the same model he used to explain organic diversity, one in which the natural order of things was altered but never wholly destroyed by constraining factors in particular environments. Lamarck assumed that fluctuations in the weather displayed certain regularities owing to the influence of the sun and the moon on the atmosphere. The sun acted through the production of light and through gravitation. The moon acted through gravitation only, but had an especially large effect owing to its proximity to the earth. Since the positions of the sun and moon could be predicted in advance, their influence on the earth's atmosphere was also predictable. The problem was that every anticipated influence on the atmosphere was subject to modification by the pre-existing state of the atmosphere at the particular place and time in question.[43]

Lamarck was disappointed with the little support that his own meteorological studies, and meteorological studies in general, received from the scientific societies of his day. He attributed this neglect of meteorology in part to the general opinion that atmospheric phenomena were too complex and too irregular to be lawful.[44] But he also saw something more sinister in his contemporaries' failure to acknowledge the value of what he was doing. Once again it seemed to him that persons in power, unwilling to admit the merit of ideas that were not their own, were conspiring to undermine his scientific efforts.[45]

In contemplating how the simplest forms of life might be affected by adverse conditions and how subtle fluids, continually circulating at every point of the earth's surface, influenced organic activity, Lamarck's biological thought did come in contact with issues of a meteorological nature. But his meteorological theorizing in and of itself had no direct bearing on his biology. The major extrabiological influence on Lamarck's biological thought is found in his geology.

LAMARCK'S GEOLOGY

In February 1799 Lamarck presented to the first class of the institute a memoir entitled "On fossils and the influence of the movement of the waters, considered as proofs of the continual displacement of the sea bed and its transport across the points of the globe's surface."[46] This memoir was never published and no manuscript of it is known to exist, but one may infer from its title that it was a preliminary version of Lamarck's *Hydrogéologie* (1802). This inference is supported by a brief contemporary report on activities at the institute, which gives some further details on the contents of Lamarck's memoir:

> An attentive examination of the fossil shells that are found in all the habitable parts of the earth has been for citizen Lamarck a nearly incontestable indication of the presence of the sea for a great number of centuries on each point of the globe's surface now raised above the level of the waters. This author thinks that no sudden catastrophe whatsoever could have brought these remains of marine animals and deposited them in the state in which we now see them. According to him, the bed of the seas owes its origin and conservation to the oscillatory movements of the sea waters, perpetually maintained by the influence of the moon; and this bed experiences a continual though insensible displacement, causing it successively to cross over the globe's entire surface.[47]

The above account does not include all the subjects that Lamarck was to discuss in the *Hydrogéologie,* but it does include most of the major ones. Of the ideas not mentioned, the most important for Lamarck's theorizing was his idea of the organic origin of minerals, which he had already advanced more than a decade earlier. The major features of his geological theory thus appear to have taken shape by February 1799, a little more than one year before he first presented his evolutionary views. Indeed, as early as 1797, in his *Mémoires de physique et d'histoire naturelle,* he maintained that the sea "has successively covered the continents."[48] But at that time he was willing to "leave aside these large views, in which it is too easy and too common to lose oneself."[49] Two years later he was ready to confront these views directly in the memoir he read to the institute. Insofar as Lamarck's geological views seem to have been formed by 1799, they will be treated here as part of the body of thought he developed prior to becoming an evolutionist.

Lamarck's geological theory provides a prime example of how his general failure to specify the source of his ideas has led historians of science to an exaggerated opinion of his originality. An inspection of the writings of Buffon, Baumé, Daubenton, Lamarck's friend Bruguière, or any of a number of other late eighteenth century thinkers reveals how little, if any, of Lamarck's *Hydrogéologie* was original. The basic hypothesis Lamarck used in his *Hydrogéologie* can be found in the early geological thought of Buffon: the hypothesis of the slow but continual displacement of the ocean basins on the earth's surface as the result of the motions of the earth's waters, which results from the combined effects of the gravitational pull of the moon and the rotation of the earth. In the writings of Baumé are to be found not only the ideas of the organic origin and the mutability of minerals, but also the idea that the shaping of the earth's surface was primarily the result of organic action and water in motion.[50] A lecture that Daubenton delivered to students at the *École normale* in 1795 contained the basic views that Buffon had expressed more than forty years earlier and that Lamarck was to express in his *Hydrogéologie* seven years later. Daubenton maintained that water plays a dominant role in shaping the surface of the earth and that the oceans in the course of the earth's history have covered every point of the earth's surface, their beds being slowly displaced from the east to the west largely as the result of tidal action. On the basis of data concerning the nature of the ocean floor, the presumed results of tidal action, the geological strata evidently explicable only as the result of long-term deposition, and the distribution of fossil shells and other organic remains, Daubenton felt able to conclude "that the waters of the sea have covered the continents, and that, in the course of time, the vast ocean will cover them again, when it has left, little by little, the immense space that it occupies presently."[51]

The ideas held by Buffon, Baumé, and Daubenton illustrate how incorrect it is to assume that Lamarck's *Hydrogéologie* was a work of great originality. The major geological views Lamarck presented had been endorsed by prominent naturalists before him. Indeed, these views had been around for more than two decades. A comparison of Lamarck's thinking with a newer arrival on the scientific scene, one whom he respected highly and to whom he was apparently much indebted, is also helpful. Jean-Guillaume Bruguière was an eminent conchologist, the author of the first volume of the section of the *Encyclopédie méthodique* on the Linnean class of "worms," and the one person Lamarck referred to in his published writings as "my friend."[52]

Six years Lamarck's junior, Bruguière was a Montpellier-trained physician who had developed an avid interest in natural history during his student days. He pursued that interest until it finally cost him his life. In the years 1773–1774 he traveled as a botanist with the Kerguélen expedition around the Cape of Good Hope to the South Seas, making a great many observations on marine invertebrates in particular and bringing back many plants and animals previously unknown to European naturalists. Settling in Paris in 1781, he was entrusted by Daubenton with the preparation of the section of the *Encyclopédie méthodique* on worms. He was a founding member of both the Société Linnéenne de Paris and the Société d'Histoire Naturelle de Paris. Together with Lamarck, Olivier, Pelletier, and Haüy, he was responsible for the appearance of the *Journal d'histoire naturelle* (1792). Late in 1792 he and Olivier left Paris to explore the Ottoman Empire, Egypt, and Persia. Poor health and the hardships of the enterprise proved fatal to him. He died in Ancône in 1798 during the return trip to France.[53]

When Lamarck became professor of "the insects, worms, and microscopic animals" at the museum he inevitably found himself confronting the same zoological and geological issues that had interested Bruguière. Bruguière was the man whom Lamarck regarded, at the beginning of the 1790s, as more knowledgeable concerning the "worms"—"that part of natural history so vast but so poorly known"—than any other naturalist in Paris.[54] Early in the 1790s Lamarck and Bruguière met regularly at the home of the Danish ambassador, Hwass, to explore their mutual interest in conchology.[55] The two posed some interesting contrasts: Lamarck small and frail, Bruguière obese; Lamarck prone to speculation, Bruguière reluctant to generalize without sufficient evidence. But in their passion for conchology, in their careful work as systematists, and in at least some of their thoughts on the broader significance of their studies they were quite close to one another. Lamarck, in a memoir published in 1799, just after Bruguière's death, indicated that in his lessons at the museum, he had been embracing "the principles and views" of his friend and colleague.[56] It appears that Lamarck embraced not only some of Bruguière's ideas about animal classification but other of his ideas as well.

The particular elements of Bruguière's thought that were expressed in his writings from 1789 to 1792 and that reappear and play a significant role in Lamarck's geological writings of the early 1800s are (1) a recognition of the importance of the invertebrates as geological agents; (2) the identification of invertebrate paleontology as a major key to the understanding of the history of the earth; (3) a uniformitarian view of geological change; and (4) a reluctance to believe in the extinction of species. Some of Bruguière's observations regarding the phenomena of life in the simplest animals also found expression in Lamarck's works. Though Lamarck may not have borrowed all these ideas directly from Bruguière, it is evident that he valued Bruguière's opinions.

Bruguière, like a number of naturalists before him, concluded that the invertebrates, and in particular the so-called testacean worms, had been of immense importance in the building up of a great deal, if not all, of the calcareous geological strata of the earth. "The worms," he wrote in 1789, "are . . . intimately tied to the physical organization of our globe," and he added: "It is in comparing their fossil remains from . . . remote times with those of the species that [presently] populate the vast expanse of the seas that we may now arrive at accurate ideas on the true theory [of the globe]."[57] "Fifteen years of assiduous researches," as Bruguière wrote in 1792, had qualified him to deal with the subject of the implication of fossils for geological theory.[58] Unfortunately, his travels and his death in 1798 prevented him from spelling out his views on this subject at any length. He did specify that he subscribed to the view that "the temperature of Europe was quite different from what it is now, since the shells that lived there are found there no longer—rather they are found where there is the temperature necessary for their existence."[59] Further evidence of this climatic change was to be found, he said, in the fossil bones of marine and terrestrial vertebrates and in the imprints of plants found in the schists of coal mines.[60]

In a discussion of coal mines and the imprints of plants found in the schists of these mines, Bruguière set forth most clearly his view of geological change.[61] He had visited mines near Montpellier sometime during the period 1775–1781.[62] When reporting his observations in the *Journal d'histoire naturelle* in 1792 he maintained: "Far from attributing, with Antoine de Jussieu, the formation of these immense deposits of plants to some violent perturbation of our globe, everything in the composition of these mines presents to me the image of order and the traces of slowness."[63] The surface of the earth, he said, had been changing according to "always equal and always active causes that still produce the same effects today," and it was by "the continuous repetition over a longe period of time" of a "simple and natural cause" that the great masses of plants had been piled up in the sea.[64]

There were some differences between Bruguière's geological views and Lamarck's. Bruguière does not appear to have shared Lamarck's views on the organic origin of minerals. Lamarck did not share Bruguière's idea that the oceans were shrinking, a view held not only by Bruguière but by Buffon in his later years and by other naturalists as well. But the differences between Lamarck's geological views and Bruguière's were less striking than the similarities. In particular, at a time when many naturalists were willing to appeal to catastrophes as a means of explaining a host of geological phenomena, both Bruguière and Lamarck believed that the phenomena of geology could be accounted for by the uniformitarian operation over time of nature's day-to-day processes.

The suggestion that Lamarck's geological views were not as novel as has often been supposed does not mean that these views were of little consequence for the development of his evolutionary thought. His geological thinking provided him with at least two concepts of considerable importance for his evolutionary theory: (1) the notion of an indefinitely old earth, ancient enough to have allowed all the different forms of life to have developed successively from the simplest forms, and (2) a thoroughgoing uniformitarianism. In addition, though he did not develop this point at any length, his geological theory provided for climatic change over time at each point of the earth's surface, and climatic change, as he saw it, was one of the major stimuli of organic change.

Lamarck was bold in his language, but he was not alone when he exclaimed in his *Hydrogéologie:* "How great is the antiquity of the terrestrial globe, and how small are the ideas of those who attribute to the existence of this globe an age of six-thousand and several-hundred years from its origin up to our time."[65] He shared this view with a good number of contemporary naturalists. It was maintained by the senior naturalists of the Jardin du Roi at the time Lamarck began his botanical studies. Early in the century, Fontenelle expressed the idea that the relative brevity of human life prevents the individual from appreciating changes that only take place over long periods of time and supports the belief that the world has always been as it is at present. In the memory of those roses that last only a day, Fontenelle observed, there has always been the same gardener.[66] When Lamarck presented his thoughts on organic mutability in the early 1800s, he was criticized by naturalists who pointed out that organic forms had remained stable throughout recorded history. Lamarck responded with an analogy similar to Fontenelle's, but less gentle:

> I fancy I hear those small insects that live for only a year, that inhabit some corner of a building, . . . busy consulting tradition among themselves in order to pronounce on the age of the edifice in which they happen to be. Going back twenty-five generations in their meager history, they would decide unanimously that the building which provides their asylum is eternal, or at least that it has always existed, for they have always seen it as it is, and they never heard it said that it had a beginning.
>
> Great magnitudes, in space and time, are relative.[67]

Lamarck was not precise about how old he considered the earth to be, but he was unequivocal in his assertion that it was much older than most people supposed. "For nature," Lamarck wrote, "time is nothing, and never a difficulty. It is always at her disposal, an unlimited means with which she accomplishes the greatest and smallest things."[68] A virtually identical sentiment was expressed by Desmarest, Lamarck's colleague at the institute, at the beginning of the 1800s. Introducing some brief comments on the "impenetrable abyss of past centuries" and how the supposed six-thousand-year age of the earth was merely an uninformed prejudice, Desmarest observed: "It is time, that great worker which nothing can resist, which has through its duration insensibly disfigured all our continents, by the means of all the destructive agents it puts to work each day . . . The more one follows the traces of nature's operations with a rational and reflective examination, the more one is able to convince oneself that these immense works could have been accomplished by all the agents we see operating before our eyes."[69] Lamarck's colleague at the museum, Barthélemy Faujas de Saint-Fond, though not sharing Lamarck's uniformitarian views, also believed in the great antiquity of the earth.[70] Lavoisier, delivering to the academy in 1789 the only geological memoir of his career, spoke of the formation of deep-sea shell beds requiring "an immense succession of years and centuries."[71] Bruguière likewise believed in the great antiquity of the earth, though his published statements do not indicate just how old he believed the earth to be.

Lamarck himself described the age of the earth as "nearly incalculable"[72] and "so great that it is absolutely beyond the power of man to appreciate it in any way."[73] He did indicate in his *Hydrogéologie* that the time scale he had in mind involved thousands[74] or even millions[75] of *centuries,* and he estimated that the time it would take the beds of the oceans to make a complete revolution of the earth—which he believed had occurred at least once[76]—was roughly 900 million years.[77]

Arguing from fossil evidence, as had others before him, Lamarck asserted that the beds of the oceans had gradually displaced themselves over time and that at each point on the earth's surface the climate was subject to a "continuous though infinitely slow change."[78] He explained the displacement of the ocean beds in a way similar to Buffon (and Daubenton) and saw in this displacement the basic cause of climatic change: "As the displacement of the ocean basin produces a constantly varying inequality in the mass of the terrestrial radii, this necessarily causes the center of gravity of the globe to vary, as well as its two poles. Moreover, as it seems that this variation, irregular as it is, is not subject to any limits, it is very probable that each point of the surface of our planet may successively be subjected to all the different climates."[79]

Lamarck's increased attention to geological theory at the end of the eighteenth century coincides with the development of his evolutionary views, both chronologically and logically. Indeed, he often portrayed organic change as a consequence of geological change. When in 1802 he began publishing his researches on invertebrate paleontology, he referred his readers to his *Recherches sur l'organisation des corps vivans,* claiming he had established there "that living bodies undergo changes in their form and even in their organization, in proportion as they experience changes forced upon them in their habits, their way of life, and through external impressions upon them."[80] He added "I have shown that [living bodies] are subjected to these changes when the circumstances in which they live are considerably changed."[81] Given that Lamarck's geological theory provided for inevitable change at each point of the earth's surface, it would seem that organic change was likewise inevitable (and that it would occur not because of any mysterious *inherent* tendency to change, but rather because of the operation of physical factors depending ultimately upon gravitational mechanics).

However, the connection between organic and geological change that Lamarck drew in the 1800s was not one that he had ever made explicit in the 1790s. Though in 1802 he was prepared to state that "living bodies undergo changes in their form and even in their organization in *proportion* as they experience changes forced upon them" earlier in his career he had emphasized the "*limits* that nature has imposed . . . on the various changes that circumstances can effect upon plants."[82] What caused him to change from a belief in limited mutability to a belief that mutability is proportional to environmental change (and hence virtually unbounded over long periods of time) remains unanswered. There is no reason to suppose that geological ideas that were readily available to Lamarck from the beginning of his scientific career were sufficient to force him into a belief in evolution. In addition to the complex of geological ideas discussed above, however, there was one problem that, though broached on numerous occasions in the eighteenth century, became a burning issue in the French scientific community only in the latter half of the 1790s. This was the issue of species extinction. When Lamarck became involved in this issue, he became involved not merely as a naturalist-philosopher interested in nature's general operations, but also as an expert with the particular knowledge required to resolve some of the questions about extinction. Species extinction was the issue that brought him to the idea of species change.

There is no doubt that the ideas on chemistry, meteorology, and geology that Lamarck maintained in the 1790s had a bearing on his biological and evolutionary thought of the following decade. His ideas on the production of minerals by the successive breakdown of the complex products of organic action, on the role of subtle fluids in the life processes, and on the amount of time available in the earth's history for changes to have taken place in a uniformitarian fashion, all seem highly suggestive when viewed in the light of his later evolutionary theories. Yet, at base, his theory of organic evolution was not a necessary consequence of his thinking in these other fields. His physico-chemical system denied the possibility of the direct generation of life from non-life, a possibility that was central to the formulation of his evolutionary theory. His meteorological work was never oriented toward the study of the influences of atmospheric conditions on living things, and his thoughts on the operation of these influences were scarcely more developed in the 1800s than they were in the 1770s. Finally, the endorsement of uniformitarianism in geology does not demand the acceptance of evolution in biology. Indeed, Lamarck shared this geological view with a number of naturalists, none of whom adopted his view of biological change.

One should not pass from what has been labeled here "the preoccupations of the new professor" without drawing some general conclusions regarding Lamarck's scientific style on the eve of his development of an evolutionary theory. Lamarck's work in chemistry, geology, and meteorology displays a characteristic impatience to get at the basic generalities underlying natural phenomena. His chemical and geological treatises are filled with boldly stated views but lack carefully established empirical foundations. His meteorological writings were no different. He accompanied his meteorological theorizing with tables of observations and probabilities, but this gave his ideas only the aura of an empirical basis. His meteorology had as its basic assumption the reasonable enough idea that the earth's atmosphere, like the earth's waters, was affected by the gravitational attraction of the moon. But he was never able to demonstrate to the satisfaction of his contemporaries that the action of the moon and sun on the earth's atmosphere could be identified in his tables of observations, let alone predicted in advance. It was only in his work as a naturalist that Lamarck displayed that control of detail that his contemporaries insisted was essential to the progress of science. But if in his work as a zoologist Lamarck was able to display again those skills as a systematist that had first earned him a reputation in botany, he was not content to be a mere namer and classifier of species. From the cabinet of the naturalist he set out not simply to categorize nature's productions but to comprehend the basic phenomena of life.

CHAPTER FIVE

Invertebrate Zoology and the Inspiration of Lamarck's Evolutionary Views

According to Étienne Geoffroy Saint-Hilaire, when the professorships were parceled out at the Museum d'Histoire Naturelle, Lamarck was left with the subject no one else cared about: that of invertebrate zoology.[1] There is no evidence, however, that Lamarck was particularly displeased with his fate. He knew that, with the exception of the insects, the lower animals had been the subject of relatively little systematic investigation, but he was prepared to argue that his new province of study was as significant intellectually as any other part of nature's domain. As it turned out, the study of the lower animals proved even more enlightening to Lamarck than he initially anticipated. In contemplating what the less well known part of the animal kingdom revealed about nature's processes, he was led to formulate a general science that he called "biology," which had as one of its central elements a bold new explanation of the origin of living things.

When Lamarck began his study of the "insects, worms, and microscopic animals" he was able to draw on the work of two of his colleagues, J.-G. Bruguière and G. A. Olivier, who had recently departed on an expedition to Persia and the Ottoman Empire. The year after he assumed his new teaching duties he was also able to draw on the work of Georges Cuvier, whose arrival in Paris in 1795 had an immediate and profound impact on the natural history practiced there. Cuvier's skill and experience as a comparative anatomist lent special force to what some Parisian naturalists had recognized in principle but few had realized in practice, namely, that the study of animal organization held the key to the problems of classification and physiology alike.

By the 1780s most of the major French naturalists had dropped the traditional, tripartite division of nature in favor of the distinction between brute bodies on the one hand and organized bodies on the other. Even in the field of conchology, where there was an inevitable tendency to classify shells according to their external features rather than the anatomy of the animals that inhabited them, the importance of animal structure had been acknowledged. Three of the most distinguished conchologists of the century—the Frenchmen Michel Adanson and Étienne Geoffroy and the Dane Otto F. Müller—had based their conchological systems on the features of the animal within the shell rather than the shell itself, and J.-G. Bruguière, the leading French conchologist at the beginning of the 1790s, maintained that the ideal conchological system would involve a consideration of animals and shells at the same time.[2]

With the exceptions of Daubenton and especially Vicq-d'Azyr (1748–1794), however, the French naturalists of the late eighteenth century were not noted as comparative anatomists. Indeed, when the museum began functioning in 1794, its zoologists could hardly be counted its greatest asset. Lamarck's expertise lay in botany, not invertebrate zoology. Étienne Geoffroy Saint-Hilaire, the professor of vertebrate zoology, was only twenty-two years old and had not yet developed any scientific specialty. Though Geoffroy's later career certainly bore out Daubenton's faith in him, certain Parisian naturalists had a legitimate reason to be upset by Geoffroy's appointment.[3] Lacépède had initially been destined to occupy the position Geoffroy received, but political difficulties prevented him from accepting the professorship in 1793. Lacépède, however, was not a great scientist himself. He became a professor at the museum in 1795, splitting the vertebrate zoology responsibilities with Geoffroy, but he was frankly more talented as a composer of rapturous passages about the "magnificent scenes" of nature than he was as a naturalist. Daubenton, whose anatomical work had added so much to Buffon's *Histoire naturelle,* had turned his attentions to mineralogy. Mertrud, though a skilled dissector of the human body, was old and was without experience as a comparative anatomist. With Vicq-d'Azyr dead and Bruguière and Olivier off exploring the Ottoman Empire, the way was clear for Cuvier to become the leading zoologist of Paris.

Before describing the contributions of Cuvier and Lamarck to the classification of the lower animals, mention should be made of some of the special advantages enjoyed by those who worked with the collections at the Muséum d'Histoire Naturelle. Cuvier profited immensely from the materials available to him at the museum. Lamarck, who in addition to having access to the museum collections, amassed two very important collections of his own (one of plants and the other of shells), frequently referred to the importance of working with large collections.[4] As the museum's collections grew in the 1790s and 1800s, it became the best place in Europe from which to write a *Règne animal* (Cuvier's major systematic work) or a *Histoire naturelle des animaux sans vertèbres* (Lamarck's major systematic work).[5]

At the beginning of the 1790s the zoological collections at the Cabinet du Roi were satisfactory if not excellent. The resources of the new museum were increased dramatically in the next few years when the "Commission of the Sciences and the Arts" accompanying the French armies in Europe appropriated objects of natural history in conquered areas and sent them back to France (the commission included two professors of the museum, Thouin and Faujas de Saint-Fond). A report in the *Décade philosophique* in 1795 announced a special acquisition on the commission's part: "Spa, Aix-la-Chapelle, Liège, Cologne, Bonn, Coblentz have paid their contribution... But it is in Holland that the harvest has been ample and rich. The Stathouder at the Hague had a superb cabinet of natural history."[6] Three small ships, the report stated, were already on their way to France loaded entirely with boxes of specimens. The "representative of the French people" had explained the fate of the Stathouder collection to the States General of the Dutch Republic in the following terms:

> The French people, forced to fight for their liberty and to carry their arms to foreign countries, have not neglected in the course of this war to extend their learning and to increase the domain of the sciences...
>
> It is thus that, in no way contradicting the greatness and the generosity of their national character, the French have conquered less for themselves than for the human race, the happiness and liberty of which are attached to the propagation of useful knowledge...
>
> The Stathouder shrouded in obscurity these august monuments of your history. It is quite just to return them to public veneration.[7]

Public access to the natural history collections at the museum in Paris may have been freer than to the Stathouder's collection at

Excitement over the interpretation of fossils reached a new peak at the end of the eighteenth century. Pictured here is the removal of a great fossil reptile jaw from the caverns of the mountain of Saint-Pierre de Maestricht in Holland. First discovered in 1780, the fossil was the prized possession of a local cleric in 1795 when French troops appropriated it for the Muséum d'Histoire Naturelle in Paris. From Faujas de Saint-Fond, *Histoire naturelle de la montagne de Saint-Pierre de Maestricht* (1799). By permission of the Houghton Library, Harvard University.

the Hague. There is no question, however, that those who profited immediately from the French appropriation of the Stathouder's collection were the professors at the museum. When the first 150 cases from the Stathouder cabinet arrived in Paris, the zoological specimens were taken out and displayed in the amphitheater of the museum. As the *Magasin encyclopédique* reported:

> One notices there all those animals so well described by Bodaert, Allaemand, Camper, and Vosmaer; several of those brought from India and pictured in the large works of Seba, Martyn, etc. One sees there a rhinoceros skeleton, two skins and a skeleton of a giraffe, a skeleton of an unknown tailless monkey, a satyr monkey, and a host of new mammals which will soon be known through the efforts of the knowledgeable and tireless Professor Geoffroy. The ornithology is likewise very precious . . . , but the quantity of insects taken from the Cape at Batavia in Surinam is prodigious: it is they which served in the large works of Merian, Cramer, and Sepp.[8]

A number of the more important memoirs written by the zoologists at the museum during its first years depended upon specimens from the Stathouder collection. These memoirs included Cuvier's study of the living and fossil species of elephants, in which he first sketched out his ideas on extinction, Cuvier's and Geoffroy's study of the orang-outangs, and Lamarck's study of the distinctive features of the cuttlefish, squid, and octupus, his first substantial zoological publication after becoming professor at the museum.[9]

The professors of the museum also benefited considerably from French expeditions of exploration, such as that led by Captain Baudin to the South Seas (1800–1804), which epitomized both the hazards and the rewards of such enterprises.[10] The venture began with a large complement of scientists, but ill health early in the voyage forced a number of these to go no further than Ile-de-France (Mauritius), and others decided to give up the expedition at the same time. Remaining with the expedition at that point were one astronomer, one geographer, three zoologists, one botanist, two mineralogists, three gardeners, and two artists. Six of these thirteen died, as did Baudin himself, before the expedition returned to France. Though these losses were dismaying, the contributions of the expedition to natural history were considerable, as A.-L. de Jussieu hastened to point out.[11] New plants and animals had been discovered, and there was hope that some of these might be naturalized and prove to be of economic value. New specimens were add[illegible]e museum's collections. The zoological collections al[illegible]414 new specimens representing 3,872 species, [illegible]e previously unknown. The great majority of [illegible]were invertebrates, and hence were Lamarck's [illegible]y. Considering the strength of the collections at the [illegible]um, one is led to conclude that the professorships were positions not only of prestige but of considerable scientific advantage as well.

The revolution that Cuvier effected in the classification of the lower animals, however, was something he accomplished when he first came to Paris. He had had little time to work with the museum collections. He had, however, had the experience of studying marine life at first hand, in the years he served as a tutor to a noble family in Normandy. At any rate, on May 10, 1795, he read a memoir to the Société d'Histoire Naturelle de Paris entitled "On the internal and external structure and on the affinities of the animals that have been named worms."[12] His expressed goal was "to make known more exactly the nature and the true relations of the white-blooded animals, by reducing to general principles what is known about their structure and properties."[13] Speaking of "the admirable fecundity of the principle of the subordination of characters" and the "beautiful laws it leads us to," he divided the "white-blooded animals" into six classes, having paid special attention first to the circulatory systems and then to the nervous systems of the animals in question. The classes, arranged according to their "different degrees of organic perfection," were the mollusks, the crustaceans, the insects, the worms, the echinoderms, and the zoophytes. He attached special importance to the placement of the mollusks at the head of the white-blooded animals. He maintained that the complexity of the mollusks' organization showed that this was the position assigned to them by nature.[14]

Lamarck was convinced that structural complexity was the critical feature for determining the "natural order" of nature's productions, and he welcomed Cuvier's anatomical observations. In 1796, in the introductory lecture of his course, he announced his intention to follow, in very large measure, the divisions of the animal scale Cuvier had introduced.[15] The following year, and again in 1799, Lamarck suggested that Cuvier deserved the basic credit for having reformed invertebrate classification.[16] Cuvier, in the meantime, mentioned Lamarck as one of the naturalists to whom he was indebted,[17] and blithely observed that the naturalists of the day were "busy clearing the vast field of nature together." "Provided a discovery be made," Cuvier wrote, "it matters little to the naturalists who among them or their friends attaches his name to it."[18]

The friendly relations that Cuvier and Lamarck apparently enjoyed in the late 1790s did not last. Neither Cuvier nor Lamarck was as oblivious to credit as Cuvier airily suggested. When Cuvier wrote a review of Lamarck's *Système des animaux sans vertèbres* in 1801, he took care to begin the review with a description of what he himself had accomplished back in 1795.[19] He allowed that he himself had not provided a complete system of the "white-blooded animals" insofar as he had not dealt with the genera and the species. He then paid Lamarck a backhanded compliment: "[Lamarck's new book] is the best proof that no one is more capable than he of occupying himself successfully with *what remains to be done* to complete the methodical arrangement of this part of the animal kingdom" (italics added).[20]

Lamarck was not pleased with the follow-the-leader role Cuvier was prepared to assign him, and in his later writings he attempted to claim more credit for himself in the early reform of invertebrate classification. In his *Philosophie zoologique* (1809) he maintained that he had introduced several important classificatory innovations during his first year of teaching, 1794.[21] He began, he said, by dividing all the animals into two major groups: the vertebrates (*les animaux à vertèbres*) and the invertebrates (*les animaux sans vertèbres*). He accepted the classes of vertebrates indicated by Linnaeus—the mammals, birds, amphibians (and reptiles), and fish—but he was not content to divide the remaining animals into the two classes of insects and worms proposed by Linnaeus and his successors. "No one could deny that the class of worms of Linnaeus was a kind of chaos in which very different things were united," Lamarck observed, "but the authority of this savant was of such great weight for naturalists that no one dared change this monstrous class."[22] Bruguière had divided the Linnaean class of worms into six orders "according to their apparent organization, together with their faculties and the places they live," arranged as follows: (1) the infusorians, (2) the intestinal worms, (3) the mollusks, (4) the echinoderms, (5) the testaceans, and (6) the zoophytes.[23] Lamarck claimed that in his own lessons, he divided the invertebrates into five classes and arranged them in a series of decreasing complexity of organization: mollusks, insects, worms, echinoderms, and polyps. "These classes," Lamarck explained, "were composed at that time of some of the orders that Bruguière had presented in his distribution of worms (the arrangement of which, however, I did not adopt) and the class of insects as circumscribed by Linnaeus."[24] Cuvier's contribution in 1795, Lamarck suggested, consisted of providing the "decisive proofs" for things Lamarck himself had already expounded in his own lessons: "The change that I made . . . by feeling the inconvenience of the accepted Linnaean distribution was perfectly consolidated by Cuvier through the exposition of the most positive facts, among which some, in truth, were already known, but had not yet attracted our attention in Paris."[25] Lamarck's feeling that Cuvier had received too much of the credit for the reform of the classification of the invertebrates was hinted at again in 1812 when he observed bitterly that experience had taught him "what can result from a slowness in recording, through publication, the observations that one makes in public lessons."[26]

Beyond Lamarck's own assertions there is no evidence that he divided the invertebrates into five classes before Cuvier's arrival in Paris. There is some evidence, however, in support of Lamarck's claim that very early in his lessons he separated the mollusks from the insects and worms, making the mollusks a distinct class, which he placed below the fish and above the insects and worms on a linear scale. In the *Magasin encyclopédique* of 1796, Antoine-Nicolas Duchesne presented a table of "the inhabitants of the waters, distributed according to the various classes of organized beings."[27] Duchesne remarked that his table was similar to one he had presented earlier, "with the exception of the last classes containing the invertebrate animals, of which a division into three, following the views of Professor Lamarck, seems to me to be required by the most recent anatomical observations."[28] In Duchesne's table the invertebrates (*invertebroses*) are divided into mollusks, insects, and worms, with the mollusks heading the list. The date of Duchesne's article does not allow certainty on this matter, but it seems likely that "the views of Professor Lamarck" to which Duchesne referred were views expressed in 1794 or 1795, since as of 1796 Lamarck was basically following the divisions suggested by Cuvier.

From 1796 to 1799 Lamarck continued to work on refining the classification of the invertebrates, offering with special satisfaction arrangements of the orders of insects (1797) and the genera of shells (1799).[29] What he apparently did not alter significantly during this period was the introductory discourse that he delivered to his students at the museum. A brief contemporary account of how Lamarck opened his course in 1796 describes his introductory comments as follows: "[Lamarck] treated the necessity of devoting oneself to the study of the insects and worms. He reported a host of curious observations that make this study very attractive. He then set himself to prove that man can make use of a great number of these beings and that there are others that he ought to attempt to destroy. Citizen Lamarck then treated the most natural division to use for the insects and the worms."[30] This description of Lamarck's introductory discourse of 1796 fits perfectly Lamarck's introductory discourse of 1799, the manuscript of which still exists.

Perhaps the most striking thing about the discourse with which Lamarck opened his course throughout the 1790s is its lack of originality in at least two of its three major sections. Indeed, it appears that in arguing for the importance of studying invertebrate zoology, Lamarck did little more in his first year as professor than slap together statements by Bruguière and Olivier. He delivered the same statements, largely unchanged, through 1799. In the third section, on the classification of the invertebrates, he was perhaps more original, but, as he admitted there, his opinions on invertebrate classification were only slightly different from Cuvier's. Lamarck's heavy borrowing from Bruguière and Olivier is not all that surprising if one recognizes Lamarck's esteem for those two men and his preoccupation with other subjects for much of the 1790s. That he completely rewrote his introductory discourse between 1799 and 1800 seems strong evidence that in those years his views underwent a major restructuring.

The manuscript of the introductory discourse Lamarck delivered in 1799 opens with a broad statement of purpose: "I propose to make known to you a particular series of animals, . . . that which alone comprises more species than all the others in the same kingdom taken together; that which at the same time is the most fertile in marvels, in striking peculiarities, in interest of all kinds in regard to us; and however that which perhaps is still the least well known."[31] Having identified the animals in question as the "invertebrates" or "white-blooded" animals, Lamarck launched into a discussion of the attractiveness of the study of these animals. This discussion was taken almost word for word from Bruguière's introduction to the volume on worms of the *Encyclopédie méthodique*.[32] A few paragraphs later Lamarck began a discussion of invertebrates harmful to man and what man could do to diminish the unwanted effects of these beings. This section was borrowed virtually word for word from an article in the *Journal d'histoire naturelle* by Olivier.[33] In neither case did Lamarck acknowledge his debt. Where Lamarck's own ideas emerged, if at all, was in his discussion of the subject "so attractive for the naturalist, so worthy of a *philosophe observateur,*" namely, the study of the natural beings "according to their true affinities."[34]

In his study of the invertebrates, as in his botanical work, Lamarck's thinking about the true affinities among nature's productions was guided by the idea of serial arrangement. As he explained to his students, he intended to direct their attention to "the truly graduated series [formed by the invertebrates], and the points of repose or lines of separation that nature herself seems to have formed in this immense series," and "the principal facts of organization that necessarily characterize the classes, orders, natural families, and often the genera of this interesting series."[35]

Cuvier eventually broke radically with the notion of a scale of being in the animal kingdom, announcing that the animal kingdom was represented by four distinct plans of organization that could not be arranged hierarchically. In the 1790s, however, Cuvier arranged the "white-blooded animals" in a linear sequence according to their complexity. The idea of serial arrangement had come under some criticism in the last decades of the eighteenth century, however, primarily with respect to the notion that the different species could be placed in a line. Among the critics of serial arrangement was Daubenton, the patriarch of the museum. Daubenton was not as resolutely opposed to theories and systems as some writers have suggested, but there was one system that he was particularly eager to combat: Bonnet's scale of nature.[36] In 1782 he turned his attention to that "great question of natural history," the question of "whether Nature passes from one species to another by successive nuances; whether all the species of her production could be arranged on the same line, so that each species would have closer resemblances [*rapports*] with those surrounding it than with the others; or whether this order, instead of being continuous, would be interrupted by gaps between the species that would not have the appropriate characters to form a kind of connection between them."[37] Daubenton admitted that "most often Nature passes from one species to another by differences so slight that they form nuances that are nearly imperceptible,"[38] but he concluded that "the differences between Nature's productions do not succeed one another in a direct line—to the contrary they follow several oblique lines."[39]

In the lessons that Daubenton delivered at the École Normale in 1795 he continued his critique of the idea that nature's productions could be arranged in a straight line, attempting, as he put it, "to destroy [this idea] by proofs that will permit no reply."[40] While most of his remarks were directed against Bonnet, Daubenton also mentioned "the author of the *Flore française*" as a scientist who had suggested arranging the productions of nature linearly.[41] Concerning his own studies on the quadrupeds, Daubenton reported:

> I combined [fifteen distinctive characteristics of the quadrupeds] in every way. I considered them according to all sorts of relations. However I disovered no continuous sequence from one species of quadruped to the others, no direct order, no progression of characters that might indicate the place that each species ought to occupy relative to the others in order to be arranged along the same line. But I have seen clearly several lines joined to each other at different points.[42]

Lamarck, who was setting to work on the classification of the invertebrates at this time, was sensitive to such criticism. He had given up his early thought of arranging the plant species in a single line. He never suggested the possibility of such an arrangement for the animal species. While recognizing that reticular arrangements of nature's productions were becoming increasingly popular, however, he maintained that the most natural arrangement in each of the two organic kingdoms was still basically a linear one. As he indicated in 1801, he supposed that the "error" of those who had suggested reticular arrangements of nature's productions would fade away as the knowledge of the organization of living things became more profound and more general.[43]

The consideration of the organization of the different invertebrates, Lamarck indicated in his discourse of 1799, led naturally to the examination of their different faculties. "We have learned already," he told his students, that in proportion as the perfection of organization diminishes or seems to diminish, that is to say, in proportion as the organization is simplified, the faculties of animality acquire in general a much greater breadth."[44] "In the insects, even more so in the worms, and especially in the polyps," he explained, "the faculties of animality are in truth less numerous but much more extensive than in the mammals. Irritability is much greater there, the faculty of regenerating parts is easier, and that of multiplying the individuals is much more considerable."[45] As he put it early in the discourse, borrowing the words of Bruguière, "[The imagination] is even frightened by some of their faculties."[46]

Lamarck's discourse of 1799 does not indicate that as of the spring of that year he had come to believe either that species were mutable or that nature had produced all the different forms of life successively, from the simplest to the most complex. Amid the statements that he borrowed from Bruguière, however, are passages in his own words that seem, at least in retrospect, to show him on the verge of a change in his thinking, if he had not already made it (italics have been added to identify the words Lamarck added to Bruguière's original statement):

> If one considers the *prodigious* number of *invertebrate* animals *and especially those* which the naturalists have designated under the names of *insects,* worms, *and polyps;* if one observes in some the simplicity of their organization and in others the singular apparatus they present; finally if one reflects on the so-varied modes of their propagation and their regenerations, whether natural or artificial; *one cannot help but admire nature's infinite resources in the variety of means she employs to diversify, multiply, and conserve her productions, that is, the species and the kinds that constitute them.*[47]

To a modern-day evolutionist, Lamarck's discussion of nature "diversifying" and "multiplying" her productions may seem like a clear endorsement of an evolutionary viewpoint. This interpretation is invalid, however. In its context Lamarck's phrase about nature "diversifying" her productions refers only to the variety of different forms of organization displayed by the invertebrates, while nature "multiplying" her productions is surely no more than a reference to the reproduction of individuals, not a reference to what is now called speciation.[48]

That Lamarck had not yet arrived in 1799 at an idea that was central to his evolutionary theorizing is borne out by the next passage (in which italics have again been added to Lamarck's words to distinguish them from Bruguière's):

> The waters are populated *in a way* with animated molecules that are endowed with organs as perfect *for assuring their existence* as those of the largest animals, *though the organization of the latter is more com-*

plex. Thus these animated molecules reproduce themselves as constantly and even with a facility and promptness that is much superior [to the larger animals]; and we can be sure that they hold a rank in nature just as unequivocal, though in general less well known.[49]

The point of most importance in this passage from the standpoint of understanding the development of Lamarck's thinking is his assertion that the organization of the simplest animals was sufficient to assure their existence, and that the rank these organisms held in nature was consequently "unequivocal." Nature had taken care, in other words, to *conserve* her productions (as he said in the passage cited previously). In 1799 he does not appear to have felt that the life of the simplest animals was especially precarious, at least not relative to their reproductive abilities. The following year, however, he was prepared to say that the simplest forms of life could be entirely destroyed by intemperate conditions, *and then generated anew from inanimate matter once favorable conditions returned.* He was prepared to believe in direct or spontaneous generation, an idea that directly contradicted his earlier chemical thinking about the inability of brute matter to produce life. Between 1799 and 1800, apparently, Lamarck's thoughts on the nature of life in the simplest of all animals, the "animated molecules," developed significantly, and the impact on the rest of his thoughts was dramatic.

IN DISCUSSING the actual inspiration of Lamarck's evolutionary views, two different ideas need to be distinguished from one another. One is the idea of species mutability. The other is the idea that nature began with the simplest animals and then from them produced successively the more complex animals. To endorse the latter idea, Lamarck had to come to believe in spontaneous generation and in the idea that the graduated scale of complexity he was finding in the animal kingdom represented "nature's true order" in a phylogenetic as well as a formal sense. Lamarck's early thoughts on species mutability could simply have been deduced from this more general view of organic change, but it seems in fact that they were not. It appears instead that they were inspired by a particular issue in late eighteenth-century natural history.

Reading Lamarck's major theoretical works in biology—his *Philosophie zoologique* and the long introduction to his *Histoire naturelle des animaux sans vertèbres,* one may forget (or more likely never realize) just where Lamarck's greatest expertise as an invertebrate zoologist lay. Similarly, looking over Lamarck's record of publication in the early 1800s, one would very likely not guess what it was that his contemporaries were eagerly awaiting from him in 1799. As indicated earlier in this study, however, when he was chosen as professor of the "insects, worms, and microscopic animals" at the museum, Lamarck's only real claim to competence in that field derived from his work as a conchologist—a collector and systematizer of shells. In 1799 it was not a *Philosophie zoologique* that his contemporaries awaited from him, nor a *Recherches sur l'organisation des crops vivans,* nor a *Hydrogéologie,* nor even his *Système des animaux sans vertèbres*—at least not as such. The work to which they looked forward was to be entitled *Elémens de conchyliologie.* It had been promised them by Lamarck late in 1798 when he presented to the first class of the institute an "introduction to a new classification of shells."[50] He indicated at that time that the work was well advanced and that he hoped to publish it soon. As he explained his purpose, "It seemed to me indispensable to give to the public, and especially to the students following my lessons at the museum, some *elements of conchology,* a work in which I plan to present a concise exposition of principles relative to the study of shells, and to their distinction in families, genera, and species."[51]

Contemporary interest in Lamarck's proposed work was not simply the interest that any systematic exposition of a part of nature's productions would have aroused. To be sure, such a work had special attraction because shell collecting was quite a fashion in Paris at that time.[52] The particular interest in Lamarck's work, however, had another, more significant basis. Conchology was seen as having a special contribution to make in resolving the most exciting issue in natural history at the end of the century: namely, the reality and the extent of species extinction in the course of the earth's history.

The possibility of species extinction had occurred to naturalists well before the 1790s. It had been impressed upon them by the existence of fossils with no known living analogs. It was not until the 1790s, however, that the reality of extinction began to be demonstrated in a convincing manner. The scientist primarily responsible for this was Cuvier. In 1796, at the first public session of the institute, he used his special talents as a comparative anatomist to argue that there were two species of elephants currently in existence, not one, and that these species were not only different from each other but also from the species of elephants whose bones had been preserved as fossils.[53] He noted further that although the earth contained the fossil remains of all sorts of animals, it appeared that none of these animals were still in existence. The obvious conclusion, he felt, was that "the organized beings that exist at present have replaced other [beings] that some catastrophe totally destroyed."[54] Having raised the issue, however, he displayed characteristic caution and refrained from grappling with it:

> But what then was this primitive earth where all the beings differed from those that have succeeded them? What nature was this that was not subject to man's dominion? And what revolution was capable of destroying it to the point of leaving as trace of it only some half-decomposed bones? But it is not our duty to engage ourselves in the vast field that these questions present. Let hardier philosophers undertake it. Modest anatomy, limited to detailed examination, to the meticulous comparison of objects submitted to its eyes and to its scalpel, will content itself with the honor of having opened this new route to the genius who will dare to travel along it.[55]

In support of the idea of a "revolution" that had brought about the disappearance of an entire fauna, Cuvier claimed, among other things, that "several learned conchologists maintain that none of the shellfish now living in the sea are to be found among the numerous petrifactions of which the continents are full."[56] Significantly, this claim appears only in the manuscript version of the memoir read at the institute. In the published versions of the memoir, the reference to conchologists was omitted. Historians of science, unaware that Cuvier's claim about support from conchology was ever made, have overlooked the importance of conchology for the debate on species extinction at the turn of the century. To the conchologist in Paris in the late 1790s, however, the importance of his subject for the debate about extinction was clear. Furthermore, from the men with the greatest expertise in that subject, support for Cuvier's claim was not forthcoming.

Barthélemy Faujas de Saint-Fond, professor of geology at the museum, was one of those in 1799 who awaited eagerly the publication of Lamarck's proposed *Conchyliologie*. As Faujas explained his interest at that time: "One of the principle causes that has up to the present hindered the advancement of the natural history of fossils, and consequently geology, is the incertitude which has always reigned over the knowledge of shells because of the lack of a good methodical work on this subject."[57] Though Lamarck's *Elémens de conchyliologie* never appeared as such, descriptions that would have gone into that work found their place in Lamarck's *Système des animaux sans vertèbres,* which was published in 1801. In the *Système* Lamarck added 18 more genera to the 126 genera of shells he had already named in his 1799 sketch of shell classification. Faujas rushed quickly to the *Système*—in fact he worked from page proofs that Lamarck lent him—in order to tackle the question that was of such great interest to the naturalists of the day: "Do fossil or petrified shells in fact exist that have analogs, living presently in this or that sea, which are incontestably recognized as belonging to the same species? Or are there sufficiently pronounced differences [between these forms] not to allow one to consider them as [being of the same species], as some people maintain?"[58] Faujas was able right away to identify forty-one species of fossil shells with known living analogs, and he supposed that as others used Lamarck's work the list would grow.[59]

Lamarck himself, of course, was not a man content to cede to others the means by which the "large facts" of nature could be perceived. He knew well the broader implications of his conchological studies. In 1798 he had announced his intention to include in his *Conchyliologie* an indication "of the living or marine shells that are truly analogous to certain of our fossil shells."[60] Any errors in determining these analogs, he indicated, "would be very prejudicial to our researches in this interesting part of natural history . . . Only by the accuracy of the determinations of those of our living or marine shellfish that are analogous to the fossils of our continents will we be able to obtain solid and well-founded deductions on several important points concerning the theory of our globe."[61]

Lamarck's expertise as a conchologist brought him directly to the issue of whether species had become extinct through large-scale "revolutions" in the earth's past. He was the one to whom French naturalists looked to provide the necessary information on the similarities and differences between fossil and living shells. It was the twin specters of extinction and global catastrophe that drove him to a belief in the mutability of species.

Neither the idea of species extinction or the notion of an extraordinary geological upheaval was compatible with the operations of nature as Lamarck conceived them. Catastrophism he scorned as unscientific: "A universal upheaval that necessarily regularizes nothing, mixing and dispersing all, is a very convenient means for those naturalists who wish to explain everything and do not take the trouble to observe and study nature's way in regard to her productions and all that constitutes her domain."[62] His own view of nature's operations was wholly uniformitarian. One could account for the different features of the earth's surface, he believed, by ordinary processes of change working over long periods of time. As for his rejection of the idea of species extinction, this was based on two sorts of arguments: (1) nature's balance was such that species were not lost; and (2) a natural mechanism by which many species might have become extinct was inconceivable.

When he analyzed the concept of nature, Lamarck concluded that nature was a nonintelligent order of things, "which acts only by necessity, and which can execute only what it does execute."[63] Lamarck believed, however, in an essentially well-ordered universe, and it was natural for him to speak at least metaphorically of "nature's wisdom." In the lecture to his students in which he first outlined his evolutionary ideas, he spoke of the "wise precautions of nature" through which "everything remains in order": "No species predominates to the point of bringing about the ruin of another, except perhaps in the most complex classes, where the multiplication of individuals is slow and difficult . . . In general the species are preserved."[64] Nine years later, in his *Philosophie zoologique,* he wrote: "It remains a question for me, to know whether the means employed by nature to assure the conservation of the species or the races have been so insufficient that entire races are now annihilated or lost."[65] The statement was rhetorical. Lamarck had no doubt that nature's means were adequate to protect her productions.

Of the species "said to be lost" Lamarck would go no further than admit that a few large quadrupeds *might* have become extinct. If such extinction had taken place, he indicated, it could only have been caused by man himself. The loss of even a few species was not to be ascribed to a failing on nature's part. In his fullest statement of his understanding of nature's balances, Lamarck explained how the general equilibrium was maintained. The fecundity of the smaller species was such, he said, that they "would render the globe uninhabitable for the others, if nature had not set a limit to their prodigious multiplication. But since they serve as prey to a multitude of other animals, since the length of their life is very limited, and since drops in temperature cause them to perish, their quantity is always maintained in just proportions for the conservation of their own races and others." The larger and stronger animals, he explained, might be thought to pose a threat to the other species, "but their races devour one another, and they only multiply slowly and a small number at a time, which conserves in their regard the sort of equilibrium which ought to exist."[66] Lamarck's view of the "wise precautions" through which "everything in the established order is conserved" inspired one of his more eloquent passages:

> The perpetual changes and renewals observable in this order are maintained within unsurpassable limits; the races of living beings all subsist despite their variations; the progress acquired in the perfecting of organization is not lost; all that seems to be disorder, destruction, and anomaly returns without cease into the general order, and even contributes to it; and everywhere and always, the wish of the sublime Author of nature and of all that exists is invariably executed.[67]

It has been suggested that Lamarck's reluctance to admit species extinction was due both to his acceptance of traditional ideas concerning nature's balance and to his "conception of life as a principle of order and organization."[68] Lamarck himself, however, never drew any connection between his thoughts on the balance of nature and his often-expressed view of life as a force operating against brute nature. If there was such a connection, it was never explicit. There is a simpler, less abstract way of explaining Lamarck's reluctance to admit species extinction.

Lamarck rejected the idea of species extinction because, for all but a few organisms, he could not imagine a natural mechanism for extinction. In "the most complex classes, where the multiplication of individuals is slow and difficult," it was at least possible for Lamarck to imagine that man's own destructive powers might have caused the extinction of a few species. At the opposite end of the animal scale, among the simplest animals of all, it was possible for Lamarck to imagine that a sudden climatic shift could wipe out a species that lay completely naked before the elements. (This idea, as will be indicated below, seems to have led directly to his idea of spontaneous generation.) But for the vast range of animals between the two extremes of the animal scale, how could their extinction be imagined? What could bring about the destruction of any of those insects that multiplied so prolifically? What could cause the demise of those many marine invertebrates with which Lamarck was familiar, when their ocean bottom habitat sheltered them not only from man's depredations but also from inclement climatic conditions? Lamarck could conceive of no natural means by which such organisms could ever have become extinct, and he refused to attribute extinction to some extraordinary catastrophe in the earth's history.

Lamarck's conclusion concerning the possibility of the extinction of prolific marine organisms was essentially the same as that of his late friend Bruguière. Bruguière had been quite interested in invertebrate paleontology and had intended to publish a "general history of the fossil shells of the kingdom," maintaining that the examination of fossil and living shells was the key to the true theory of the earth.[69] Confronted with the problem of why many known fossil forms had no known living counterparts, his explanation was straightforward: the living counterparts of these forms were simply yet to be discovered. In 1792 he estimated that man, having many parts of the world still to explore, knew probably no more than a fifth of the species of shells in existence.[70] He justified taking the time to describe fossil shells by saying: "I must not neglect describing the fossil shells since it is proven [!] that all these shells have their marine analogs and many of these were known for a long time in the fossil state before individuals of the same species were found living in the sea."[71]

Bruguière offered the genus *Ceritus,* first discovered in living form on one of Cook's voyages, as an example of a common fossil with living analogs, and he claimed that he knew other examples as well.[72] He advanced the view that the reason why such well-known fossil forms as the ammonities and the belemnites had never been found except in a fossilized state was that they lived only in the greatest depths of the sea. Ammonities must still be alive, he maintained in 1789, because there was no conceivable way they could have been destroyed: "What causes could be adduced [for the extinction of these creatures] when it is virtually demonstrated that the temperature must be equal in the ocean at a great depth, and that if it is proved that [these creatures] once lived there, they must, by this reason alone, be found there still."[73]

The point that Bruguière made in 1789 proved to be a telling one for Lamarck a decade or so later. Holding firmly to his uniformitarian view of nature's processes, Lamarck refused to accept a major catastrophe as a means of explaining extinction. That left him, however, with no way at all of explaining extinction on a broad scale. He concluded, therefore, that such extinction had not occurred. But if one were to deny, as Lamarck did, that species had ever been lost on a large scale, only two ways remained to explain why many of the fossils with which scientists were familiar had no known living counterparts. Lamarck used them both. The first was to suppose, as Bruguière had done before him, that living counterparts of these fossils did in fact exist in parts of the globe that scientists had yet to explore. In Lamarck's words: "There are yet so many portions of the surface of the globe where we have not penetrated, so many others that capable observers have crossed only in passing, and so many others still, as the different parts of the depths of the seas, for which we have few means of identifying the animals that live there, that these different places could well be hiding the species with which we are not familiar."[74]

For the invertebrate zoologist in the late eighteenth century this position, as Bruguière's example indicates, was still tenable. Uncharted shores and the depths of the seas continued to yield living representatives of forms of life that had first been known only in the fossil state. The prime example in the early 1800s was the discovery of a trigonian clam, which Lamarck reported in the *Annales* of the museum.[75] Frequently, when he described a species known only as a fossil, Lamarck would add a sentence to the effect that "the living analogue of this species is not *yet* known." Such a position was more difficult to maintain in the case of the vertebrates than in that of the invertebrates. Nonetheless, Lamarck was not alone in his reluctance to accept Cuvier's thoughts about broad-scale extinction.

Lamarck's second explanation of why living analogs of fossil forms were rarely found was of special consequence for the further development of his thought. It was a deduction that Lamarck, by his unwillingness to accept the extinction of species on anything but a very limited level, was naturally led to make. It meant the negation of a view that he had held for a long time, but more fundamentally, it meant the salvaging of a view that was of greater significance to him as naturalist-philosopher. To retain his view of the well-balanced, uniformly operating processes of nature, he gave up his belief in species fixity.

Lamarck explained the problem in the *Philosophie zoologique,* as follows: "If many fossil shells display differences that do not allow us, according to accepted opinion, to regard them as analogs of neighboring species with which we are familiar, does it necessarily follow that these shells belong to species that are really extinct [*perdues*]? Why, moreover, should they be extinct, since man could not have brought about their destruction?"[76] His resolution of the problem was a simple one: "Would it not be possible, on the contrary, that the individual fossils in question might belong to species still existing, but which have since changed, and have given rise to the species now living that we find to differ slightly from them?"[77]

Lamarck did not have to argue from first principles alone. His conchological studies provided him with enough examples of living counterparts of fossil forms to discount the idea of a universal catastrophe. Furthermore, though he found many fossil forms with no known living counterparts, for many of these he could point to living species of the same genus that resembled the fossil species quite closely. Unwilling to admit that some major catastrophe might have caused extinction, unable to conceive of a natural mechanism by which extinction might have taken place, but granting that there were differences between fossil and living forms that were unlikely to be explained simply by the discovery of living analogs of the fossils, Lamarck was led quite naturally to a belief in species mutability.

In summary, the explanation of how Lamarck arrived at his belief in species mutability is fairly straightforward. Three possible explanations of the observed differences between fossil and living forms were recognized at the turn of the century. As Cuvier saw the problem in 1801, the question at issue was "just how far the catastrophe that preceded the formation of our present continents went," and to answer the question one had to determine "whether the species that existed then have been entirely destroyed, or whether they have been only modified in their form, or whether they have been simply transported from one climate to another."[78]

Cuvier initially adopted the first of these solutions—wholesale extinction. Bruguière had favored the idea of migration. Lamarck relied in part on the idea of migration but also embraced the idea that in the course of the earth's history organic forms have been modified.

This explains the inspiration of Lamarck's belief in species mutability, but it is not the whole story of the inspiration of Lamarck's evolutionary views. Lamarck's understanding of evolution was not simply (or even primarily) an extension of the idea of species mutability that arose from his refusal to admit species extinction. It was one thing for him to maintain that "although many fossil shells are different from all the marine shells known, that in no way proves that the species of these shells are destroyed, but only that these species have changed in the course of time, and that presently they have different forms than those of the individuals whose fossil remains we find."[79] It was quite another thing for him to maintain that nature had begun by producing the very simplest forms of life and then from these had successively produced all the others. Buffon's case shows that it was possible for an eighteenth-century naturalist to accept a certain amount of organic mutability without adopting the sort of evolutionary view Lamarck eventually presented, and Buffon's example was repeated by a number of Parisian naturalists at the turn of the century. Lacépède (professor of reptiles and fish at the museum), Delamétherie (editor of the *Journal de physique et d'histoire naturelle*), and Denys de Montfort (an avid conchologist and friend of Faujas de Saint-Fond) all sketched out, or at least hinted at, naturalistic explanations of the origin of living things, in which the modern forms of life were derived from a greater or lesser number of spontaneously generated forms of different degrees of complexity.[80] Of those of Lamarck's contemporaries who were willing to admit the possibility of significant organic mutability, only Faujas de Saint-Fond wrote of the possibility of nature "proceeding progressively from the simple to the complex."[81] This Faujas did only once, late in 1799. Given his close contact with Lamarck at the time, Faujas' statement is perhaps an indicaton that he and Lamarck had discussed the idea of progressive evolution prior to Lamarck's public espousal of it.

Lamarck's idea, that nature proceeded from the simple to the complex in producing all living things, came not from his consideration of fossils, but from new thoughts he had just developed on the nature and origin of life, together with old thoughts he had long maintained on the natural way to arrange nature's productions. As indicated in the previous chapter, between 1794 and 1797 Lamarck's statements on the nature of life changed in an important way: he dropped entirely his vitalistic terminology of 1794. In 1797, however, he still treated nature and the "power of life" as two opposing forces—one destroying and the other creating and renewing. Life, he maintained, was always the product of life. "It has been rightly said," he observed, "that everything that has life comes from an egg."[82] Three years later, in contrast, he was prepared to express a belief that the "sketches of animalization" at the "unknown end of the animal scale" were organisms that nature "forms and multiplies with . . . facility in favorable circumstances."[83] In the following years he specified the physical conditions necessary for the production of life from non-life. Though he still spoke in 1800 of an "immense hiatus" that could be said to exist between brute matter and living things, he had come to believe in spontaneous generation or, as he preferred to call it, "direct" generation.

Lamarck never recounted how he came to believe in spontaneous generation. In 1802 he argued that "for living bodies to be truly the productions of nature, nature must have had and must still have the ability to produce some of them directly."[84] But he did not explain what led him to take literally the common phrase "productions of nature." His thinking about the nature of life in the simplest animal forms suggests, however, what it was that convinced him that the direct generation of these organisms was possible.

Lamarck maintained that if one really wanted to know what constitutes the essence of life, one had to examine the simplest beings in which the phenomena of life exhibited themselves. The simplest forms of life were, in his view, genuinely simple. The monad, "the most imperfect and the most simple of the known animals," he described later as "nothing but a point that is gelatinous and transparent, but contractile."[85] Given this idea of the extreme simplicity of the organisms at the very beginning of the animal scale, it appeared to him that the "excitatory cause of organic movement" in these simplest creatures and the "sole cause of their conservation" had to be external to them."[86] Botanist turned zoologist, he had been led to reflect upon the fundamental similarities and differences between plants and animals. In 1797 he had maintained that life in plants was "an activity provoked by an external cause."[87] He seems to have held this view of plant physiology since the 1770s, when he wrote of the way daily temperature changes caused the motion of plant fluids. By 1800 he had decided that the simplest animals were as dependent on external influences for their vitality as plants were. Two subtle fluids, heat and electricity, were in Lamarck's view the stimuli necessary for life in the simplest beings.[88] Viewing life as an organic movement, and viewing subtle fluids as responsible for organic movement within the simplest organisms, it was natural enough for him to suppose that the very simplest organisms, the "mere sketches of animality," could be formed when conditions were favorable and destroyed when conditions were unfavorable.[89]

Lamarck's *Mémoires de physique et d'histoire naturelle* of 1797 and manuscript notes from 1798 and 1799 indicate that in those years he was pondering the problem of generation. His interest in both plant and animal physiology had led him to consider the mechanisms of such phenomena as fertilization, germination, and incubation. Fertilization, he had come to believe, "is nothing other than an act that establishes a particular disposition in the interior parts of the gelatinous body that is fertilized—a disposition without which the individual would never be able to receive and enjoy life."[90] It was just a step more to believe that, as he put it in 1802, "nature herself imitates her process of fertilization in another state of things, without having need of the concourse or products of any pre-existent organization."[91] Heat and electricity, he proposed, could do to certain materials under the appropriate conditions just what the "*subtle vapor* of fertilizing materials" did to embryos, namely, organize them and make them ready for life.[92]

While it is possible that conceiving how spontaneous generation *could* come about was what made Lamarck believe that spontaneous *did* come about, there is some reason to suspect that what inspired Lamarck's belief in spontaneous generation in the first place was a consideration of the problem of extinction as it applied to the very simplest forms of life. As indicated above, he rejected the idea of extinction for virtually all the species of the animal kingdom because he could not imagine natural means by which these species could have become extinct. For the very simplest forms of life, however, for the so-called monad, the sketch of animality that was but a gelatinous, contractile point, it was easy to conceive of a mechanism of extinction—the unexpected arrival of a "rigorous season."[93] As both a meteorologist concerned with forecasting major shifts in the weather and a Frenchman who had lived through the 1780s and 1790s, Lamarck had no difficulty imagining such an occurrence. From what he know of the simplest forms of life, he supposed these forms were much too frail to cope with adverse conditions. His problem thus lay in explaining the reappearance of these forms after they had apparently all been destroyed. Since they had no organs for sexual generation, it was obvious to Lamarck they could produce no eggs and thus had to reproduce instead through fission or budding. What was unknown to him was the encysted state in which simple forms survive difficult circumstances. Not knowing this, he naturally concluded that the reappearance of the simplest forms of life occurred when, under appropriate conditions, they were generated directly from inanimate matter.[94] This explanation of the inspiration of Lamarck's belief in spontaneous generation is strongly supported by the fact that from the first his endorsement of the idea that the simplest animals could be spontaneously generated under favorable conditions was connected with the complementary assumption that the same animals could be destroyed when conditions were unfavorable.

Lamarck did not expect his ideas on direct generation to meet with immediate or great approval.[95] He insisted, however, that he was justified in broaching the subject: "To try, as a naturalist, to look for what may be the *origin of living things* and how they have been formed is a condemnable temerity only in the eyes of the vulgar and ignorant, and not in those of he who is wise enough *not to assign to the supreme power, creator of all nature, the mode that it must have followed* in bringing everything into existence."[96] Lamarck thus attempted to put himself on the side of the rational seekers after truth as opposed to the narrow-minded adherents of the Scriptures. What some of Lamarck's fellow scientists must have felt, however, was that although Lamarck was not wrong in broaching the subject of spontaneous generation, his nonexperimental approach to the issue had no hope of being conclusive.

The step Lamarck had taken in coming to believe in spontaneous generation was critical for the development of his view of evolution. Others before him had believed in spontaneous generation without believing in evolution. The former belief does not necessitate the latter. But once the simplest forms of life—the forms in which the essence of life was most clearly displayed—had become in Lamarck's view literally the productions of nature, he could not but believe that nature's means were adequate to produce all the other forms of life as well. Given his long-held thoughts on the general linear arrangement of nature's productions, his views at the turn of the century on the mutability of specific forms, and his tendency to see simple patterns underlying complex phenomena of all sorts, it becomes possible to appreciate the feeling he expressed in 1802 that "once the difficult step [of admitting spontaneous generation] is made, no important obstacle stands in the way of our being able to recognize the origin and order of the different productions of nature."[97]

Lamarck's new beliefs in species mutability and spontaneous generation were the two crucial additions to his thinking in the period 1799 to 1800. Though the first of these beliefs was directly related to his consideration of the question of extinction and the second may have been related to the question of extinction as well, it is evident that he adopted these beliefs willingly, not only as responses to the question of extinction but also because of their usefulness to him in defending his old view of the natural way to arrange nature's productions.

As indicated in Chapter Two, the idea that within each of nature's kingdoms her productions could be arranged linearly according to their complexity of organization was the fundamental assumption of Lamarck's classificatory work. Though in the late eighteenth century a number of naturalists adopted alternative models of the natural order of nature's productions, Lamarck held to his view of a linear scale of organic complexity. Cuvier's anatomical researches of the 1790s appeared to Lamarck to provide further evidence of the correctness of his position, and Lamarck was pleased to announce to his students at the museum in 1800 that the invertebrates "show us even better than the [other animals] this surprising degradation in the composition of organization.[98] In 1803, in his last botanical work, Lamarck remarked that the "real gradation in the organization of living things" displayed itself "in an eminent manner" among the animals. This gradation, he admitted, was "very little evident" among the plants, but it existed there no less than in the animals.[99]

The basic assumption of Lamarck's evolutionary theory was that the scale of organic complexity was to be interpreted in phylogenetic terms. As he put it in 1803, "this real gradation in the organization of living things must necessarily be regarded as the result of the true *marche* of nature."[100] Lamarck had spoken of nature's *marche,* or way of proceeding, much earlier in his career. As of 1800, however, the phrase *marche de la nature* took on a special meaning in his writings. He used it to indicate the course nature had followed in bringing all the different forms of life into being.

By 1800 Lamarck had acknowledged that neither the species nor the genera could be arranged linearly in such a way that each form in the series would have the greatest affinities with the two forms on either side of it. But this, in his opinion, was no reason to give up the notion of serial arrangement altogether. He maintained that the general "masses" of organization, as he called them—the classes or large families—could be arranged in a single series of increasing complexity. The genera and especially the species formed lateral ramifications from this series in many places. The extremities of these ramifications represented "truly isolated points." But the series itself, Lamarck insisted, surely existed.[101]

The importance Lamarck's belief in the scale of nature had for his thoughts on organic change cannot be insisted upon too strongly. With his belief in "a real gradation in the organization of living things," and his belief in the spontaneous generation of the simplest (and only the simplest) forms of life, it made sense for Lamarck to conceive of organic mutability in terms of *progressive* development. As he urged his students in 1802:

> Ascend from the simplest to the most complex; leave from the simplest animalcule and go up along the scale to the animal richest in organization and facilities; conserve everywhere the order of relation in the masses; then you will have hold of the true thread that ties together all of nature's productions, you will have a just idea of her *marche,* and you will be convinced that the simplest of her living productions have successively given rise to all the others.[102]

Lamarck's thoughts on organic change were only presented sketchily in the introductory discourse he delivered to his students in 1800. How much of this sketchiness was due to the nature of the occasion and how much was due to the state of development of his thinking is impossible to say. The discourse was a wide-ranging treatment of the importance of studying invertebrate zoology, not an exposition of a theory of organic change. As he remarked concerning this discourse when he published it in 1801 at the beginning of his *Système des animaux sans vertebres:* "I have left a glimpse there of some important and philosophical views that the nature and the limits of this work have not permitted me to develop, but which I propose to take up again elsewhere with the details necessary to make known their foundation and with certain explanations that will prevent them from being abused."[103]

When Lamarck did set forth his "philosophical views" at further length (in his *Recherches sur l'organisation des corps vivans* of 1802), the most important addition to what he had said in 1800 was the way he accounted for the graded complexity of organization in the animal scale. In 1800 he was not explicit about the causes of organic development. What he claimed then was that for all the classes, orders, genera, and species in existence he could show "that the conformation of individuals and their parts, that their organs, their faculties, &c. &c. are entirely the result of the circumstances in which the race of each species has found itself subjected by nature."[104] Though he repeated this claim in 1802, his thinking by then had developed a step further, primarily reflecting his concern with the broad features of the animal scale. He maintained then and thereafter that the general animal series was the result of the tendency of "organic movement" in living things to make their organization increasingly complex, while the lateral ramifications from the general series—the ramifications formed by certain species and genera—were the result of the particular circumstances living things had been subjected to. With this development his evolutionary theory took on its essential shape.

CHAPTER SIX

Lamarck's Theory of Evolution

In 1815, setting forth what was to be the last major exposition of his biological and evolutionary views, Lamarck felt able to say that "on the source of the existence, the manner of being, the faculties, the variations, and the phenomena of organization of the different animals" he had presented "a truly general theory, linked everywhere in its parts, always consistent in its principles, and applicable to all the known data."[1] He had explained—to his own satisfaction, at least—how nature first created the simplest forms of life, how the organization of animals had become increasingly complex, how the higher animal faculties emerged with the increasing "perfection" of organization, and how the influence of particular "circumstances" had led to special habits and structures in different animals. Though he did not address the subject in 1815, in his earlier biological treatises of 1802 and 1809 he had even suggested how man himself might have arisen from a "less perfect" animal.

Lamarck had indeed constructed a *general* theory dealing with a host of biological phenomena (half a century later Charles Darwin would restrict his subject much more severely when he wrote his *Origin of Species*), but he had not presented his views with such clarity and consistency as to exempt them from misinterpretation. The development of biology since his day has inevitably caused certain aspects of his thought to be overemphasized while other aspects have been forgotten, resulting in a distorted picture of what his evolutionary theory was all about. The common misunderstandings regarding his theory must not all be ascribed, however, to the preoccupations of biologists of more recent times. Lamarck's writings have their ambiguities, inconsistencies, and misleading metaphors. It was often several years between his initial presentation of a major concept and his first full explanation of it. Furthermore, his thinking on certain aspects of the evolutionary process did change over time. Even in the long, theoretical introduction to his *Histoire naturelle des animaux sans vertèbres*—the last major exposition of his views—one cannot find a complete statement of his evolutionary theory. He revised his theory significantly in a "supplement" he added to the first volume of the *Histoire naturelle,* and it was only in some of the later volumes of that work (volumes that historians of biology have consistently failed to consult) that he gave some of his most explicit statements about the mechanisms of organic change.

Though Lamarck's writings pose problems for the historian of biology wishing to reconstruct Lamarck's evolutionary thinking, these problems are not insurmountable. The basic structure of Lamarck's theory remained constant from 1802 through the last exposition of his views, and during this period he continued to address himself to a number of general themes: the role of the "power of life" in making animal organization more complex, the emergence of the higher animal faculties as organic complexity increased, the course nature followed in bringing all living things into existence, and the influence of particular environmental circumstances on nature's plan. Lamarck's thoughts on these subjects can be analyzed, as can his thoughts on such topics as the inheritance of acquired characters and the importance of felt needs in the evolution of the higher organisms (topics which are now commonly associated with his name). Also open to scrutiny are the kinds of evidence Lamarck offered in support of the notion of organic mutability and, in general, Lamarck's talent (or lack of it) as a scientific strategist.

LAMARCK'S TWO-FACTOR THEORY OF ORGANIC CHANGE

Though Lamarck's theory of organic evolution departed significantly from the other theories of organic diversity that were offered prior to 1800, it was still in its essential structure a theory that belonged very much to the late eighteenth century. It involved the sort of analysis of change found in the writings of the most important political, economic, and social theorists of the time. It also reflected the naturalist's growing appreciation that some anatomical characters had to be weighted more heavily than others in the determination of the natural affinities of living things.

Lamarck's theory of evolution provided a natural history of organic forms in a special, eighteenth-century sense. When he explained what he commonly referred to as *la marche de la nature* (he himself never used the words "evolution" or "transformism"), he treated change in much the same way it was treated in Rousseau's *Discourse on the Origin of Inequality,* Adam Smith's *Wealth of Nations,* Hume's *The Natural History of Religion,* and Malthus' *Essay on the Principle of Population.*[2] In each of these works, regardless of whether the subject was society, economics, religion, or population, one was told what the *natural* course of events would have been had it not been for constraining circumstances. Dugald Stewart called it "theoretical" or "conjectural" history (which is what Hume called "natural history" and some French writers called *histoire raisonée*):

> In most cases it is of more importance to ascertain the progress that is most simple, than the progress that is most agreeable to fact; for, paradoxical as the proposition may appear, it is certainly true that the real progress is not always the most natural. It may have been determined by particular accidents, which are not likely to occur again, and which cannot be considered as forming any part of that general provision which nature has made for the improvement of the human race.[3]

Lamarck's model of organic change took into account both the "natural" progress of organic development and the modification of this progress by constraining circumstances. Though in his earliest remarks on evolution, in 1800, he spoke only of the influence of the environment as a cause of organic change, by 1802 he was also talking about the "property of organic movement"[4] in developing organization and in multiplying organs and faculties, and by 1809, in his *Philosophie zoologique,* he was explicit in portraying the diversity of animal form as the result of two separate processes: "The state in which we now see all the animals is on the one hand the product of the increasing *composition* of organization, which tends to form a *regular gradation,* and on the other hand that of the influences of a multitude of very different circumstances that continually tend to destroy the regularity in the gradation of the increasing composition of organization."[5]

Again in 1815 his distinction between two different causes of organic diversity was clear-cut:

> It is easy to perceive that [the different organization of the various animals] comes from the action of two very different causes, one of which, though incapable of destroying the predominance of the other, nevertheless very often varies the other's results.
>
> The plan of nature's operations in regard to the production of the animals is clearly indicated by the first and predominant cause, which gives to animal life the power of progressively complicating organization —not only organization considered over all, but also each particular system of organs, successively as [nature] manages to establish them. This plan, that is, this progressive complication of organization, has really been executed by the first cause in the different animals that exist.
>
> But a cause different from the above-mentioned, a cause that is accidental and consequently variable, has thwarted here and there the execution of this plan, without however destroying it . . . This cause, in effect, has given rise to the real gaps in the series, to the finite branches that spring from [the series] at diverse points and alter its simplicity, and finally to the anomalies that may be observed among the specific organ systems of the different organizations.[6]

The "first and predominant cause" represented the "natural" course of events, the "plan of nature," what would have occurred even more clearly had it not been for the second cause, the constraining influence of particular environmental circumstances. The second cause was "accidental." But this cause had to be considered if one were to understand how living things had become diversified in actuality, not just how they might have developed ideally. A constant tension existed in Lamarck's thinking between his view of the *natural* course of organic development and his view of how things had in fact developed.

Lamarck's separation of the causes of organic diversity into a "power of life" on the one hand and the modifying influences of particular environmental circumstances on the other corresponded not only to a view of change that was common to a number of eighteenth-century thinkers concerned with "natural" development, but also to his own particular work as a systematist. By the end of the century he had come to attach special importance to the idea of the subordination of characters, an idea that had been employed with considerable success by Antoine-Laurent de Jussieu in botany and Georges Cuvier in zoology.[7] In 1797 Lamarck spoke of the subordination of characters as having given natural history a stability in its principles such as it had never had previously.[8] As he explained later in the *Philosophie zoologique,* in determining the *rapports* or relations between different forms, one had to weight some characters of the organism more heavily than others.[9] It was from the characters of the internal organization of animals—the characters most essential to life—that the principle *rapports* among the animals were ascertained. In other words, it was on the basis of the consideration of organs and organ systems that the zoologist formed the different families, orders, and classes of the animal kingdom. The external characters of animals, in contrast, were what the zoologist used to determine differences on the level of species and genera. In thinking about the diversity of organic forms and the mechanisms of evolutionary change, Lamarck was inclined to attribute the development of the most essential, internal characters of organisms to the "power of life" and the development of the least essential, external characters to the influence of particular environmental circumstances. He believed that contemporary misconceptions about the natural affinities of living things would vanish as soon as more was known about the internal organization of living things and "especially when what belongs to the influence of places of habitation and to contracted habits is distinguished from that which results from the more of less advanced progress in the composition or perfecting of organization."[11]

Lamarck's statements of his two-factor theory appear unambiguous: once life was generated in simple form at the bases of the plant and animal scales it became diversified as the result of (1) the "power of life" or "cause that tends to make organization increasingly complex" and (2) the modifying influence of particular circumstances. When he came to discuss the mechanisms of organic change in greater detail, however, he was not always consistent in his claims about their operation. Furthermore, though he characteristically presented nature or the power of life on the one hand and environmental circumstances on the other as two opposing forces, a closer look at what he meant by the "power of life" reveals that the power of life was in its origin and development inseparable from environmental influences.

One example of Lamarck's inconsistency in regard to the operation of his mechanisms of organic change—an inconsistency corresponding to the tension in his thinking between the *natural* course of development and the *actual* course of development—is to be found in the *Philosophie zoologique,* where two opposing responses are given to the question of whether or not environmental change is necessary for species change. In discussing the "degradation of organization of the animal chain" Lamarck claimed: "If the cause that incessantly tends to make organization more complex were the only one influencing the form and the organs of animals, the increasing complexity of organization would be everywhere regular in its progression."[12]

More specifically he stated: "It is evident that if nature had only brought aquatic animals into being, and if all these animals always lived in the same climate, in the same sort of water, at the same depth, etc., etc., doubtlessly then we would find in the organization of these animals a regular and continuous *gradation*."[13]

He readily acknowledged that nature worked under more varied circumstances: there were fresh waters and salt waters, still waters and running waters, waters in hot climates and waters in cold climates, shallow waters and deep waters. The races of aquatic animals had become diversified according to the particular circumstances to which they had been exposed. But his meaning in the above passage seems clear: even without the stimulus of environmental change evolutionary development would take place.

Elsewhere in the *Philosophie zoologique,* however, his position was different. In regard to the mummified animals brought back from Egypt by Étienne Geoffroy Saint-Hilaire, he maintained that the similarity between these animals and modern forms was to be expected and did not constitute a refutation of the idea of organic mutability.[14] It would have been indeed strange, he observed, if these mummified forms had not been entirely similar to the forms presently living in Egypt. Since the climate of Egypt had not changed since the time when the animals were embalmed, there was no reason to expect the descendents of these animals to have changed. Cuvier had dwelt at some length upon the fact that the ibis worshipped and embalmed by the ancient Egyptians was virtually identical to a modern species (though not the one commonly supposed to be the ancient ibis) and he had then used this fact to argue for the immutability of species.[15] To Lamarck, however, it appeared obvious that "the birds that live [in Egypt] now, finding themselves in the same circumstances in which they were then, have not been forced to change their habits."[16] "Furthermore," Lamarck observed, "who does not appreciate that the birds, which can so easily move about and choose the places that suit them, are less subject to variations in local circumstances than are many other ani-

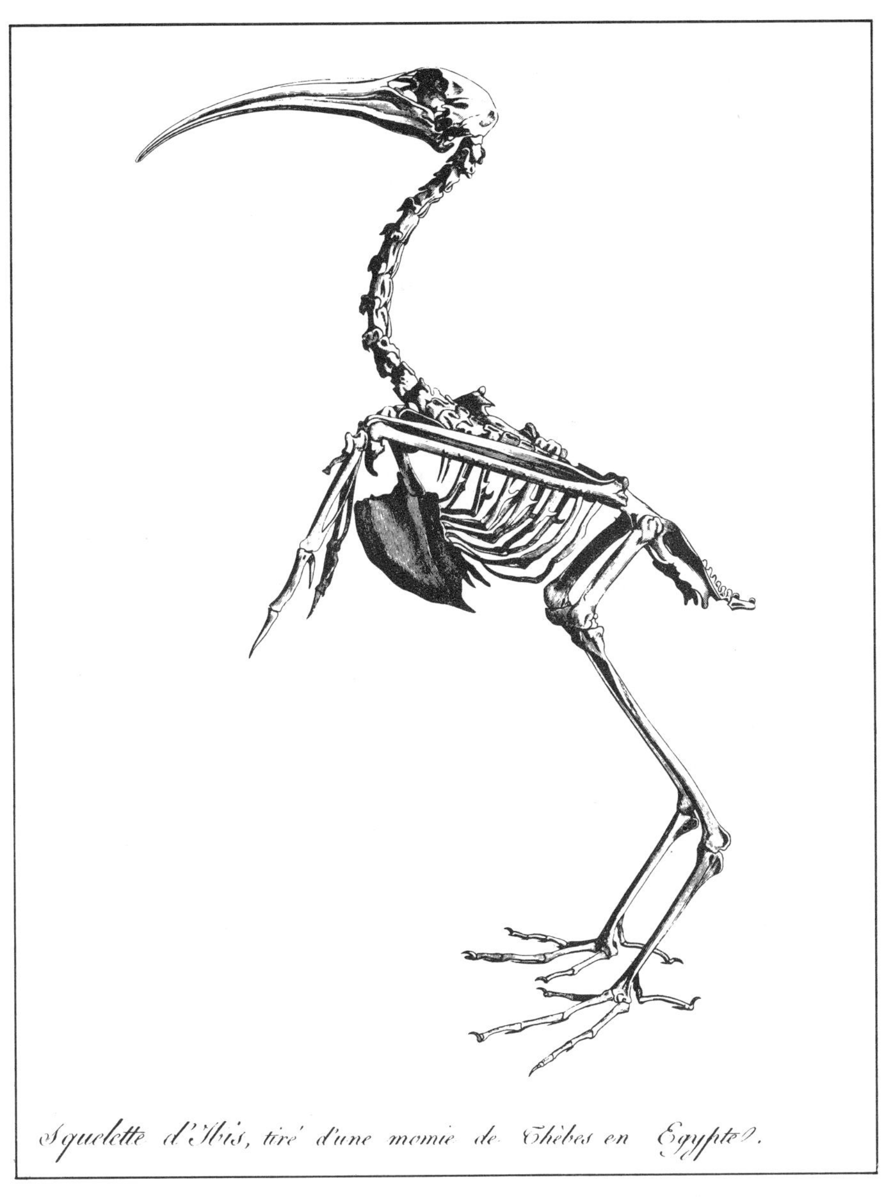

The stability of species throughout the course of recorded history was argued by Cuvier on the basis of comparisons between modern animals and animals entombed by the ancient Egyptians. Among other cases, Cuvier was able to show that the sacred Ibis of the Egyptians, as seen here in a skeleton taken from a mummy, was identical to an unknown or scarcely known modern bird, a specimen of which had come to the Muséum d'Histoire Naturelle in the Stathouder collection.

mals, and therefore less hindered in their habits."[17] Lamarck stated his view in general terms some years later: "Each [species], no doubt, is constant and always reproduces itself unchanged in the circumstances in which it normally lives; it will never change while these circumstances remain the same; that is certain, recognized, and results from the principles I have established"[18]

In speaking of "the principles I have established," Lamarck was evidently not referring to any power of life or tendency to increased complexity that operated independently of environmental change. He seems to have forgotten entirely his earlier suggestion that animals living in an unchanging aquatic environment would become regularly diversified. But if this was a genuine inconsistency on Lamarck's part, it was at least an inconsistency that followed a definite pattern. When he talked about the general series of increasing complexity formed by the animal classes, he was talking essentially about organ systems, about internal structures that were in some measure removed from the direct influence of the environment. When he talked about change at the species level, however, he was talking about changes in the external, less essential characters of organisms—the characters that presumably were most easily influenced by the environment. Thus, in the *Philosophie zoologique,* when he contrasted his own view of *species* mutability with traditional view of *species* immutability, he stressed the role of environmental circumstances in producing organic change:

The conclusion accepted up to the present: nature (or its Author) in creating the various animals foresaw all the possible kinds of circumstances in which they would have to live and gave to each species a constant organization, a form that is precise and invariable in its parts, obliging each species to live in the places and climates where it is found and to maintain there its characteristic habits.

My particular conclusion: nature, in producing successively all the species of animals, beginning with the most imperfect or most simple in order to end her work with the most perfect, has gradually made their organization more complex; and with these animals spreading generally throughout all the habitable regions of the globe, each species received from the influence of the circumstances in which it is found the habits now recognized in it and the modifications of its parts that observation shows to us . . .

My conclusion, as opposed to that accepted up to the present, supposes that, by the influence of circumstances upon habits and then by habits on the state of the parts and even on organization, each animal can receive in its parts and its organization modifications susceptible of becoming quite considerable and of having given rise to the state in which we now see all the animals.[19]

From the second paragraph of the above statement it is evident that Lamarck did have a two-factor theory of organic change in mind. From the final paragraph, however, it is easy enough to see why many persons have been led to believe that Lamarck's theory was primarily concerned with change through the inherited effects of environmentally inspired use and disuse.

THE POWER OF LIFE

Lamarck identified the primary factor of organic change to be a natural tendency toward increased complexity, which he attributed to "the power of life." With this "cause" he accounted for (and reinforced) his concept of the linear arrangement of the "masses" of animal organization. Considering his most frequent statements about this cause, it is not difficult to understand why it has been suggested that the process he had in mind was really "lawlike" rather than causal.[20] Nor is it difficult to understand how one of Lamarck's typical comments on the role of the power of life in the evolutionary process elicited from a twentieth-century scientist the comment that this "says no more than that the main movement [in evolutionary development] is an inherent characteristic of life, which certainly explains nothing."[21] But Lamarck did provide an explanation of the power of life—a very mechanical explanation based on hydraulic action and involving the solid parts of organic beings, the ponderable fluids contained within these parts, and the subtle fluids that abounded everywhere and penetrated living bodies more or less easily.

To understand what Lamarck meant by the power of life and how it accomplished all he attributed to it, it is best to begin with an examination of his thoughts on the nature of life and how the very simplest forms of life were initially produced. Despite the vitalistic ring of the phrase "the power of life," Lamarck was no vitalist. To the contrary, the whole of his biological thought was decidedly materialistic. In the 1800s, when Bichat's work was lending special strength to the vitalist position, Lamarck was adamant in his insistence that life was not a "special being" or "some sort of principle, the nature of which is unknown." He supposed to the contrary that life was "a very natural phenomenon, a physical fact."[22] The "true definition of life," as he set it forth in his *Philosophie zoologique,* was: "Life, in the parts of a body that possesses it, is an order and a state of things that permit organic movement there; and these movements, which constitute active life, result from the action of a stimulating cause that excites them."[23]

This, in Lamarck's opinion, was no arbitrary definition. Since the mid-1790s he had given considerable thought to the problem of defining life, and when he presented the above definition in 1809 it seemed to him "impossible to add or subtract a single word without destroying the integrity of the essential ideas it ought to present."[24]

Life, as Lamarck had come to view it, was a function of three interacting elements: the parts of the body that contain fluids, the fluids themselves, and the "excitatory cause" of the motions and changes that take place within the body.[25] The comparison between a living body and a watch seemed to him to be worth pursuing: the organs or supple parts of the body, together with the fluids they contained, were analogous to the wheels and machinery of movement of the watch; the exciting cause of motion within the body was analogous to the watch spring. The general features of the solid and fluid parts of the living body were, in Lamarck's opinion, well known. As for the "spring" or excitatory cause of organic movement, however, he believed that he had come upon an important idea that had previous escaped the attention of observers, namely that an *external* influence could be both the stimulus of organic movement in the simplest living things and the "sole cause of their conservation."[26]

Lamarck did not regard the simplest forms of life as extraordinary creatures. Instead, in his view, they were the forms that had to be considered if one wished to know what the essentials of life truly were. The excitatory cause of organic movement in these beings, as he saw it, was the action of those particular physical agents that were so very much a part of the scientific thought of his time:

> No doubt it would be impossible for us to recognize the excitatory cause of organic movements if the subtle, invisible, uncontainable, incessantly moving fluids that constitute it did not manifest themselves to us in a multitude of circumstances; if we did not have proofs that all the environments that living beings inhabit are perpetually filled with them, finally, if we did not know positively that the invisible fluids penetrate the masses of all these bodies more or less easily, remaining there for a greater or lesser amount of time, and that some of these are continually

in a state of agitation and expansion that gives them the faculty of distending the parts in which they are insinuated, of rarefying the fluids of the living bodies they penetrate, and of communicating to the soft parts of these same bodies an erethism or special tension that they conserve as long as they remain in a state favorable to it.[27]

Lamarck was indignant with anyone who asserted that the existence of such fluids had not been proved.[28] He admitted that it was difficult to experiment upon the "subtle, uncontainable, penetrating, and imponderable fluids," but the reality of their existence seemed to him to be incontestably demonstrated by observation.[29] When he talked to his students about the isochronous movements of the "soft radiarians," he pointed out that these movements were similar to the movements of the liquid of a thermoscope as it responded to the "caloric" escaping from his hand.[30] And this, he claimed, indicated the strong possibility that the isochronous movements of the large, soft radiarians could be considered "as the product of alternate penetrations and dissipations of the surrounding fluids that are spread throughout these bodies and exhaled from them in regular paroxysms."[31] Subtle and intangible though the fluids in question might be, Lamarck had no difficulty conceiving of their existence or visualizing their operation.

Lamarck made no claim of knowing how many different kinds of subtle fluids were involved in the "excitatory cause of organic movement" or precisely how these fluids functioned to bring about the results he attributed to them. To attempt such an explanation, he said, would be "to abuse the power of our imagination," to maintain something without the means of verifying it.[32] He believed, however, that the *biologist* could not neglect the consideration of these subtle fluids if he wished "to understand something of the phenomenon of life and to grasp the cause of the other phenomena that life can successively bring about in making the organization of animals increasingly complex."[33] He felt sure that "these invisible fluids, penetrating and accumulating and agitating incessantly within every organized body, escaping finally after having been retained there for a greater or lesser length of time, excite there movements and life, when there is an order of things allowing of such results."[34] He also felt confident in identifying two subtle fluids in particular, *caloric* and *electricity,* as playing the essential roles in the excitatory cause of life: caloric being responsible for the property of all living bodies that he called "orgasm," and electricity being responsible for the excitation of the movements and activities of the more complex animals.[35]

Lamarck's understanding of the nature of life in the simplest organisms is best revealed by his discussion of spontaneous generation in *Histoire naturelle des animaux sans vertèbres:*

We know, by observation, that the simplest organizations, whether plant or animal, are never met with except in small, very supple, very delicate, gelatinous bodies—in a word, in bodies that are frail, nearly without consistency, and for the most part transparent.

We know also that among her means of action, nature employs universal *attraction,* which tends to join together and form particular bodies; and that besides, on our globe, she employs at the same time the action of penetrating and expansive subtle fluids such as caloric, electricity, and so forth. These fluids are repellent and tend to disunite the parts of the bodies they penetrate, that is, to separate their aggregated or agglutinated molecules.

Things being thus, one conceives easily: (1) that when the small gelatinous bodies, formed easily in the waters and in humid places by the combining power, receive in their interior the expansive and repulsive fluids just cited, with which the surrounding *milieux* are constantly filled; then, the interstices of their agglutinated molecules grow larger and form utricular cavities; (2) that the most viscous parts of these gelatinous bodies, constituting, in this circumstance, the walls of the utricular cavities just mentioned, receive from the subtle and expansive fluids in question that singular tension in all their parts, that is, that kind of erethism that I have named *orgasm,* which takes part in the state of things that I have said to be essential to the existence of life in a body; (3) that once the orgasm is established in the parts of the gelatinous body in question, this body thereby immediately receives an absorbent faculty, putting it in a position to provide itself with fluids that it appropriates from outside and which fill its utricles.

In this state of things, one perceives that soon the continuous action of the subtle and expansive surrounding fluids will force the liquid in the utricles to be displaced, to open passages for itself through the feeble walls of these utricles, finally, to undergo continuous movements, susceptible of variation in speed and direction according to the circumstances.

So then, there we have the small gelatinous body we have been considering, truly organized; there it is composed of concrete containing parts, forming a very delicate cellular tissue, and of contained fluid, put unceasingly in motion by always renewed excitations from outside; in a word, there it is endowed with vital movements.

It is thus, probably, that organization was begun in the so-called spontaneous generations that nature is able to produce.[36]

For someone who considered the least complex of the known animals, the monad, as "nothing but a point that is gelatinous and transparent, but contractile," this view of spontaneous generation was not unreasonable.[37]

At the bases of both the plant and the animal scales, Lamarck presumed, tiny, transparent, unorganized masses of matter were organized and brought to life through the action of subtle fluids. The essential difference between plants and animals derived, he believed, from the fact that plants and animals were initially formed from different materials: "If the tiny mass in question is gelatinous, animal life will be established there, but if it is simply mucilaginous, then only plant life will be established there."[38] Irritability, the basic property that distinguished animals from plants, could never be established in plants because of their chemical composition.[39]

Lamarck thought the subtle fluids were responsible not only for the organization of unorganized masses of matter and the excitation of life in the simplest organic beings but also for the further development of the organization of these beings. He acknowledged that the effects of these fluids on animal form were not especially apparent in the very first "sketches of animality." The origin of the "naked infusorians," he explained, was "still too recent" for them to have become greatly diversified by the action of subtle fluids. These creatures had not been long enough subject to the conditions of life and to external circumstances to have become significantly different from the few spontaneously generated forms from which they had originally been derived.[40] In the polyps, however, the subtle fluids had begun to cause "a radiating disposition of the parts," and in the radiarians this influence was clearly pronounced. Lamarck gave one of his most explicit descriptions of the effects of these fluids on animal form in the second volume of his *Histoire naturelle des animaux sans vertèbres:*

> When, as at present, one recognizes the radiating expansibility of *caloric* and of condensed *electricity,* and one knows that all the environments inhabited by animals are more or less abundantly filled with these penetrating and expansive fluids, how can one mistake the influences these have on those animals whose parts, being still of only a feeble consistency, are consequently very supple and yield easily to the radiating expansion of these excitatory and penetrating fluids!
>
> If, in the *polyps,* these same subtle fluids have produced only a mediocre effect, who does not perceive that the very small body of each polyp has been the cause of it! But in the radiarians, where the body of each animal is much more ample and isolated, these excitatory and expansive

fluids rushing without cease through the digestive organ of these animals have evidently modified it as well as the body itself.[41]

In the coelenterates (the "soft radiarians," to use Lamarck's term), the ponderable fluids still moved very slowly, and it was to the subtle fluids that these animals owed "their vital activity, their particular movements, and even their form."[42] Lamarck explained:

> Who does not perceive, for example, that the invasion of the exciting fluids into the digestive organ of the soft radiarians, by establishing there the center of movement of the animal's own fluids, has also exercised a great influence upon the general form of the body and the disposition of its parts! Who does not perceive that, as a consequence of the divergent repulsion of these exciting fluids, the digestive organ of the *radiarians* in question had to be singularly composed and the radiating form of the body itself was the necessary result of it![43]

Lamarck supposed that in the echinoderms the direct effects of the subtle fluids on form were somewhat less than in the radiarians, owing to the increase in consistency of the bodies of the echinoderms.[44] But the ponderable fluids, once set in motion, could also shape animal form, and it was by the action of these that animal form became more and more complex. The importance of these fluids was spelled out in two propositions in Lamarck's *Recherches sur l'organisation des corps vivans* of 1802:

> That the characteristic of the movement of fluids in the supple parts of the living bodies that contain them is to trace out routes and places for deposits and outlets; to create canals and the various organs, to vary these canals and organs according to the diversity of either the movements or nature of the fluids causing them; finally, to enlarge, elongate, divide, and gradually solidify these canals and organs . . .
>
> That the state of organization in each living body has been formed little by little by the increasing influence of the movement of fluids and by the changes continually undergone there in the nature and state of these fluids through the usual succession of losses and renewals.[45]

What Lamarck meant by "the power of life" should now be reasonably clear. The power of life, as Lamarck conceived of it, resided "principally in the movement of the living body's own fluids."[46] The results of this power were displayed in the phenomena of growth and development and the regular gradation in the complexity of animal organization. But that Lamarck supposed this power to be so different from the influence of the environment is belied by his thoughts on the nature of life in the simpler invertebrates and the way these forms originated and evolved. His explanation of the production of the simpler invertebrates demonstrates that in his view the power of life was not opposed to environmental influences but, on the contrary, grew directly out of them. Only after the complexity of animal organization became sufficiently great was "the productive force of movement" internalized.[47]

FACULTIES AND ORGANIC COMPLEXITY

Lamarck's understanding of the phenomena of life displayed by the higher animals was in essence no less materialistic than his understanding of the phenomena of life displayed by the simplest invertebrates. "Every animal faculty whatsoever it be," he insisted, "is an organic phenomenon and . . . results from a system or apparatus of organs that gives rise to it."[48] Lamarck was unwilling to attribute vital properties to matter itself. Instead he claimed that vital properties were a function of organization and organic movement. An animal's organization was thus the key not only to the animal's rank in the general series of increasing complexity but also to the various faculties that the animal possessed.

In his earliest writings on the invertebrates, Lamarck maintained that as one *descended* the scale of complexity "the faculties of animality" became "less numerous but much more extensive than in the mammals. Irritability is much greater there, the faculty of regenerating parts is much easier, and that of multiplying individuals is much more considerable."[49] In his *Philosophie zoologique* he was more specific about the relation of faculties to organization. Only a few faculties, in his view, were common to all life. Each organism was capable of nourishing itself, of forming its own substance from the materials it assimilated, of growing and developing up to a certain point, and of reproducing itself.[50] Other faculties were present in some, but not all, living things. The major faculties of this sort were digestion, respiration by a special organ, action and locomotion by means of muscular organs, sensation, multiplication by sexual reproduction, circulation, and intelligence.[51] Each of these faculties depended on advances in organization beyond the organization of the simplest animals.

Digestion required an alimentary cavity, which the infusorians lacked and which did not appear until the polyps. Respiration was not a faculty of either the infusorians or the polyps. It appeared first (and then only feebly) in the radiarians, which were equipped with tracheae for the purpose.[52] The faculty of acting and moving by muscular action was clearly evident in the insects, but it was definitely not present in the infusorians or the polyps, and if it appeared at all in the radiarians it was only in the most complex of them. This faculty was dependent upon a muscular system and, prior to that, a nervous system. Sensation was an even more restricted faculty, appearing only when the nervous system was "advanced enough in its complexity to offer numerous nerves that go to a common *foyer* or *centre de rapport*."[53] Sensation first appeared in the insects, and then "in a still obscure manner."[54] Sexual reproduction required male and female reproductive organs and was of approximately the same generality as sensation in the animal kingdom. Circulation, on the other hand, was much more restricted than the five special faculties previously mentioned. It did not appear completely until the crustaceans.

Given Lamarck's thoughts on the significance of fluids in motion for organic development and on how nature could only proceed step-by-step in bringing her different productions into existence, his explanation of the appearance of the circulatory system is worth quoting at some length:

> Nature, in beginning organization in the most simple and most imperfect animals, was able to give their essential fluid only an extremely slow motion. Such is, undoubtedly, the case of the essential fluid—almost simple and very little animalized—that moves in the cellular tissue of the infusorians. But then, gradually animalizing and composing the essential fluid of the animals, she augmented its motion little by little through different means in proportion as the organization [of the animals] became more complex and more perfect.
>
> In the polyps, the essential fluid is still nearly as simple, and has not much more movement, than that of the infusorians. However, the regular form of the polyps, and especially the alimentary cavity that they possess, begins to give nature some means of activating their essential fluid a little.
>
> She probably profited from this in the radiarians, in establishing in the alimentary cavity of these animals the center of activity of their essential fluid. Indeed, the subtle, ambiant, and expansive fluids (which constitute the *excitatory cause* of the movements of these animals), penetrating principally in their alimentary cavity, through their incessantly renewed expansions have made this cavity more complex, have led to the radiating form (internal as well as external) of these same animals, and are, besides, the cause of the isochronous movements observed in the soft radiarians.

When nature succeeded in establishing muscular movement, as in the insects, and perhaps even a little before, she had then a new means of activating even a little more the motion of their *sanie* or essential fluid; but, having come to the organization of the crustaceans, this means was no longer sufficient, and it was necessary to create a particular system of organs to accelerate the essential fluids of these animals.[55]

Lamarck presumed that intelligence, like all the other special faculties, was dependent upon a particular organ. The organ in this case was the cerebral hemispheres. The addition of cerebral hemispheres to the "medullary mass" or "brain" first appeared in the fish or perhaps the highest order of mollusks.[56]

So impressed was Lamarck by the correspondence between organization and faculties that he found it appropriate, for didactic purposes at least, to divide the animal kingdom into three primary groups designated by faculties (though actually determined by organization). The first of these groups he designated as the "insensitive animals", *les animaux apathiques*. This group was comprised of the simplest of the invertebrates (those lacking any nervous system). The second group was designated the "sensitive animals," *les animaux sensibles,* and was made up of the higher invertebrates. The third group was designated the "intelligent animals," *les animaux intelligents,* and included the four vertebrate classes: the fish, reptiles, birds, and mammals.[57] For an understanding of Lamarck's evolutionary theory, most notably the way particular circumstances were capable of causing organic change, it is important to remember that he ascribed different faculties to organisms of different complexity.

THE PLAN OF NATURE

The progressive complication of animal organization was often equated in Lamarck's writings with what he referred to as nature's "plan." In the *Philosophie zoologique,* for example, he wrote: "If the vertebrates differ markedly from one another in the state of their organization, this is because nature only began the execution of her plan with respect to them in the fish, that she then advanced it further with the reptiles, that she brought it nearer to its perfection in the birds, and that she finally succeeded in terminating it completely only with the most perfect mammals."[58] As for the invertebrates, he commented: "We are convinced, upon observing their state, that to bring them successively into existence nature proceeded gradually from the most simple to the most complex. But having had as her goal the attainment of a plan of organization that would allow the greatest perfection (that of the vertebrates), a plan very different from those which she was first forced to create in order to arrive there, we perceive that among these numerous animals we ought to find not a single system of organization perfected progressively but diverse, very distinct systems, each of them necessarily deriving from the point at which each organ of primary importance began to exist."[59]

In 1812 Lamarck spoke of nature "wishing to do away with the axis" of certain polyps and accomplishing this through a series of steps.[60] In 1815 he spoke of the regular gradation in the masses of organization as nature's "real plan"—the lateral ramifications from the series being in his view the result of the thwarting of this plan by particular environmental circumstances.[61]

Lamarck evidently did not intend his references to nature's plan to be taken literally. He considered it "a genuine error to attribute to nature a purpose or any intention whatsoever in her operations," for nature, in his view, was "a particular order of things, which does not know how to will, which acts only by necessity, and which can only execute that which it does execute."[62] But if he denied that nature could really have a plan, he was nonetheless inclined to believe that nature's operations were not so complex that the attentive naturalist-philosopher could not discover the *natural* course of organic development.[63]

In his zoological studies Lamarck continually reworked the classification of the invertebrates, adding a new class to his series of increasing complexity (which he usually presented as a series of *decreasing* complexity) whenever it appeared to him that the work of the comparative anatomists had revealed the existence of another distinct plan of organization. In 1797 he recognized and arranged in a single series five classes of invertebrates: the mollusks, insects, worms, radiarians, and polyps.[64] In 1800 he increased the number of invertebrate classes to seven by adding the crustaceans and the arachnids, which he placed between the mollusks and the insects in the general series.[65] In 1802 he added the class of annelids, placing it between the mollusks and the crustaceans.[66] In 1806 he added the class of cirripeds, placing it between the mollusks and the annelids.[67] In 1809 he separated the infusorians from the polyps, making the infusorians the last class in the animal kingdom.[68] All the while he maintained his general series of increasing complexity. He repeated his classification of 1809 in a handbook he prepared for his course in 1812 and in the introduction to his *Histoire naturelle des animaux sans vertèbres* (1815).[69] Then, in a supplement to the *Histoire naturelle*'s first volume, he added a new class, the ascidians, and apparently granted the status of class to the Acephala, which he distinguished from the mollusks, and to the Epizoa, which he distinguished from the worms.[70] He quickly revised his designation of the first two of these new classes. In his final system of classification, presented in the main text of the *Histoire naturelle,* the tunicates appeared instead of the ascidians, and the Conchifera appeared instead of the Acephala. These two classes were placed between the radiarians and the mollusks. The class of Epizoa was placed between the worms and the insects.

What is most significant about the 1815 supplement to the *Histoire naturelle* is not that Lamarck added several new classes there but rather that there, for the first time, he indicated that the plan of nature had not been achieved in as straightforward a fashion as he had initially supposed. Already, in 1809, he had expressed the opinion that "the animal scale begins by at least two specific branches and that, in the course of its progress, some boughs seem to terminate it in certain places."[71] In 1815 the consideration of the plans of organization of the different invertebrates led him to the conclusion that the order of formation of the invertebrates was best represented not by one series, but by two.

Lamarck had been aware before 1815 that the intestinal worms provided special difficulties for his arrangement of the animal classes in a single series of increasing complexity of organization. The simplicity of some of these parasites, together no doubt with what was then known about their life histories, had in fact led him to suspect that the simplest of the intestinal worms, like the simplest of the infusorians, were generated spontaneously. At the time the *Philosophie zoologique* appeared, he was questioning whether "direct" generations had been limited to the most imperfect forms of life at the very beginning of the plant and animal scales. It appeared possible to him that "at the beginning of *certain separate branches* of these scales" spontaneous generations might occur.[72] Candidates for spontanous generation among the animals included not only the simplest infusorians and the intestinal worms but also "certain vermin that cause skin diseases." Candidates for spontaneous generation among the plants included certain molds, mushrooms, and lichens.[73] He did not commit himself to any of these speculations in the main text of the *Philosophie zoologique,* maintaining only that it was certain that direct generations occurred at the very base of each kingdom. In the section of additions appended to the *Philosophie zoologique,* however (the section in which he spoke of the animal series beginning by at least two separate branches), he accepted the spontaneous generation of intestinal worms. In 1812 he entertained the possibility that not only the simplest worms but also the simplest arachnids were generated spontaneously.[74] Of the various reasons in Lamarck's day to believe that intestinal worms were generated spontaneously, Lamarck, true to his classificatory principles, chose to cite the evidence from comparative anatomy: "The great disparity in organization displayed among the animals belonging to this class of worms attests . . . that the most imperfect of these animals are due to spontaneous generation."[75]

"The great disparity in organization" among the intestinal worms naturally caused Lamarck considerable difficulty when he attempted to place this class among the others in a single series of increasing complexity. The same problem arose with even greater force in 1815 when first Savigny and then Lesueur and Desmarest began announcing their anatomical discoveries on the affinities between certain animals that for the most part had been regarded as polyps and certain other animals that had generally been classed with the mollusks.[76] On the basis of the observations of these men, Lamarck felt it necessary not only to establish a new class (the one he designated first as the ascidians and later as the tunicates), but to acknowledge that the order nature had followed in forming all the different kinds of life had not been simple. It was necessary, he concluded, "to distinguish the unique and simple series that we are forced to form to facilitate our studies of the animals from the real or actual order of production of these beings, an order subjected to causes that have modified its simplicity."[77]

The work of Savigny, Lesueur, and Desmarest had caused Lamarck to identify a "singular plan of organization," which, "although more or less varied depending on the genera and species, [is] very different from the other plans of organization that characterize the animals of the other classes of invertebrates."[78] The plan of organization of this new class, Lamarck recognized immediately, was not one that allowed it to be placed without difficulty in a single series of increasing complexity. He found, though, that if he split

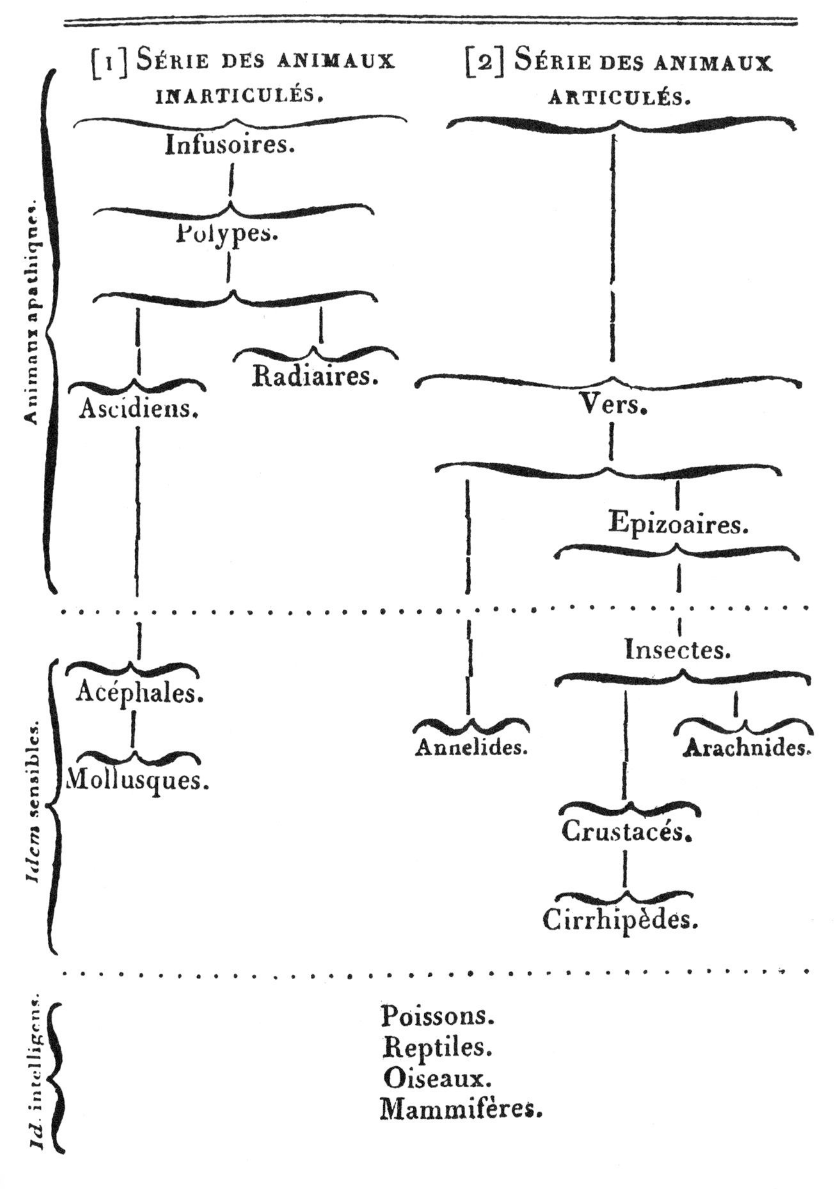

The order of production of living things, as presented by Lamarck in his *Histoire naturelle des animaux sans vertèbres* of 1815.

the invertebrates into two branches, each branch "then presented, in the relations of the objects included in it, more union between these objects and natural transitions everywhere."[79] He therefore "began to believe in the reality of these two branches."[80] In the supplement to the first volume of the *Histoire naturelle* he set forth his views schematically in a table entitled "Presumed order of formation of the animals, presenting two separate, branching series".[81] He allowed that he still considered it appropriate for didactic purposes "to have a general distribution of the animals, [a] distribution that can give us an idea of the totality of these beings, of their great diversity, of that of the particular plans of organization observable among them." But he had come to believe that it was "similarly appropriate to form another distribution that will be divided nearly to its base into two large, principal branches, in order to give us a just idea of the order of production of these beings, and to acknowledge, in each of these branches, how the laws of organization have successively introduced particular, more complex plans and the organs that come together there."[82] He explained how nature was able to form infusorians in some instances and intestinal worms in others in the same way he had explained why plants were produced in some instances and animals in others: nature produced different kinds of life when working with different materials. Thus, "A rather long time after the institution of the infusorians and the polyps, [nature] began the establishment of a new series, that of the worms, with the aid of the particular materials found inside animals already in existence. With these materials she formed spontaneous generations that are the source of the intestinal worms."[83] The single scale of increasing complexity, which he had cherished for so long, had finally been acknowledged to be inadequate in representing the true *marche de la nature*.

THE SECOND FACTOR OF ORGANIC CHANGE

The reason Lamarck gave in 1815 for nature's plan not being as simple as he had initially supposed was that "accidental causes have necessarily modified [the plan] here and there."[84] Nature was unable to form a single series because "the circumstances in which she has been forced to operate have really made her produce at least two."[85] The second cause of organic change had thwarted the tendency of the power of life to increase organic complexity in a regular fashion. In discussing how this second cause of organic change operated, Lamarck set forth the views for which he is most widely remembered today.

When Lamarck spoke of "a multitude of circumstances... that tend continually to destroy the regularity in the gradation of the growing complexity of organization," he was defining circumstances broadly.[86] "The principal circumstances [nature employs in bringing into existence all her productions]," he had explained as early as 1800, "arise from the influence of the climates, from the variations in temperature of the atmosphere and of all the surrounding *milieux*, from the diversities of places, from... habits, movements, actions, finally from the means of living, conserving oneself, defending oneself, reproducing, &c. &c."[87] In 1800, having enumerated the principal "circumstances" at nature's disposal, he described the mechanism of organic change as follows: "Now, as a result of these diverse influences, the faculties extend and strengthen themselves through use and diversify themselves through new habits long conserved; and imperceptibly the conformation, the consistency, in a word the nature and the state of the parts as well as the organs partake of the consequences of all these influences, conserving and propagating themselves through generation."[88] Lamarck gave several examples of how the habits of certain birds had determined the structure of their feet. He then asserted confidently:

> I could pass in review here all the classes, all the orders, all the genera and the species of animals that exist, and show that the structure of the individuals and of their parts, that their organs, their faculties, &c. &c. are entirely the result of the circumstances to which the race of each species has found itself subjected by nature.
>
> I could prove that it is not at all the form either of the body or its parts that gives rise to habits, to the way of life of the animals, but that to the contrary it is the habits, the way of life and all the influential circumstances that have with time established the form of the bodies and the parts of animals. With new forms, new faculties have been acquired, and little by little nature has arrived at the state where we see her now.[89]

Lamarck repeated these claims in his *Recherches sur l'organisation des corps vivans* of 1802, but by then he was attributing the basic features of the animal scale more to the general effects of "organic movement" than to particular circumstances.[90] External circumstances were merely responsible for departures from the regular gradation in the animal scale.[91] External circumstances accounted for the fact that species, unlike the general "masses" of organization, could not be arranged linearly but rather formed "lateral ramifications" around the masses, the extremities of these ramifications being "truly isolated points."[92] Later in his writings circumstances regained a place of importance in the formation of classes as well as species.

In his explanations of how circumstances brought about organic change, Lamarck stressed the importance of the use and disuse of organs and took as a basic assumption the idea that is now most frequently associated with his name: the idea of the inheritance of acquired characters. These views were spelled out in his *Philosophie zoologique* in the form of two laws:

> *First law:* In every animal that has not reached the end of its development, the more frequent and sustained use of any organ will strengthen this organ little by little, develop it, enlarge it, and give to it a power proportionate to the duration of this use; while the constant disuse of such an organ will insensibly weaken it, deteriorate it, progressively diminish its faculties, and finally cause it to disappear.
>
> *Second law:* All that nature has caused individuals to gain or lose by the influence of the circumstances to which their race has been exposed for a long time, and, consequently, by the influence of a predominant use or constant disuse of an organ or part, is conserved through generation in the new individuals descending from them, provided that these acquired changes are common to the two sexes or to those which have produced these new individuals.[93]

The ideas that use and disuse are important in organic development and that acquired characters can be inherited were not originated by Lamarck. He himself remarked in 1803 that "the law of the effects of exercise on life has been grasped for a long time by observers attentive to the phenomena of organization,"[94] while in 1815 he noted that the "law of nature by which new individuals receive all that has been acquired in organization during the lifetime of their parents is so true, so striking, so much attested by the facts, that there is no observer who has been unable to convince himself of its reality."[95] What Lamarck prided himself on was his perception of the broad significance of these laws. Thus in 1815, writing on the importance of the law of use and disuse and "the light it has shed on the causes that have brought about the astonishing diversity of the animals" he indicated that he prized more "having been the first to recognize and determine [this law] than . . . having formed some classes, some orders, many genera, and a great many species in occupying myself with the art of classification, an art that is almost the sole object of the studies of other zoologists."[96]

To appreciate the essentially mechanistic nature of Lamarck's thoughts about organic change one must consider again Lamarck's thoughts on animal organization and on how the various faculties are integrally related to complexity of organization. He looked upon the simplest animals (the "insensitive" animals, or animals lacking a nervous system) as being in much the same situation as plants. The causes of activity of these animals, he supposed, were external to them.[97] He referred to the insensitive animals as "totally passive machines."[98] By contrast, in the "sensitive" animals (the invertebrates with a nervous system), the "productive force of movement" was internalized. The primary cause of the behavior of these animals was *instinct,* which was the product of a certain *sentiment intérieur,* or inner feeling. This *sentiment intérieur* was not, however, something that introduced any true spontaneity into animal life. The *sentiment intérieur* was something that functioned in a wholly mechanical fashion. As for the intelligent animals, the vertebrates, Lamarck supposed that they could be motivated by ideas in addition to instinct but that in actuality only man and a few of the most highly developed animals were capable of acting voluntarily, and this they managed to do only rarely.[99]

Lamarck recognized fully that plants do not respond to stimuli in all the ways that the higher animals do. He did suppose, though, that plants are in much the same relationship to their surroundings as are the simplest of the invertebrates. He considered life in plants, as in animals, to be a relationship between solids and fluids, which in plants was stimulated and maintained by the action of external and variable causes, notably caloric and light.[100] And as he explained in the *Philosophie zoologique,* plants, though different from animals, are also subject to organic change:

> In plants, where there are no activities and consequently no *habits* properly speaking, great changes of circumstances nonetheless lead to great differences in the development of their parts, in such a way that these differences give birth to and develop certain of these parts, while causing several others to weaken and disappear. But here [in contrast to in animals] everything is the result of changes occurring in the plant's nutrition, in its absorptions and transpirations, in the amount of caloric, light, air and humidity it habitually receives; finally in the superiority that certain of its vital movements may acquire over others.[101]

Lamarck mentioned how *Ranunculus aquatilis* had one form when it grew underwater and another form when it was exposed to the air.[102] He also observed:

> If a seed from one of the meadow grasses . . . is transported to an elevated place that is dry, arid, stony, and much exposed to the wind, and is then allowed to germinate there, the plant able to live in this place will always be poorly nourished; and if the individuals it produces there continue to exist in these bad circumstances, the result will be a race that is truly different from the race living in the meadow, from which, however, it will have originated. The individuals of the new race will be small and thin in their parts, and as certain of their organs will have been more developed than others, they will exhibit their own particular proportions.[103]

Lamarck did not trouble himself with explaining the fine adaptations exhibited by plants. It is evident, indeed, that his theory was incapable of doing so. The mechanisms he proposed to account for changes in plants were inadequate to account for anything as complex as the various methods of seed dispersal displayed throughout the plant kingdom. He indicated that the structures of plants are harmoniously related to the conditions of their existence, but he did not worry over the details of such relationships. The examples of change in plant form that he offered involved relatively crude changes induced by altered environmental conditions.

Animals, Lamarck maintained, were in a different situation than plants, for animals had habits. In the case of the simplest animals, however, habits were "only the effect of a cause outside of the animal, a cause that acts mechanically upon it, and that necessitates its movements, whatever they be."[104] Changes in these animals were produced by the subtle and expansive fluids of the environment. Habits, Lamarck explained, arose from the accumulated effects of the motion of these fluids: "These subtle fluids, . . . according to the diversity of circumstances, . . . open different routes for themselves in the interior of the animals in question; and, once traced by repeated passings, these routes become the immediate causes of a constant likeness in the actions and the nature of the movements of the individuals of each race."[105] Habits in the most simple animals were thus the result of two things: "the invasions and the dissipations of the subtle fluids from outside" and the "particular sketch, in the organization of each species, of the routes that the subtle fluids were first forced to take and which they then always follow necessarily."[106]

The simplest invertebrates, the "insensitive" ones, did not, in Lamarck's opinion, develop their habits in response to felt needs, for they still lacked the faculty of sensation. As their organization became more complex, though, the invertebrates became sensitive, and the influence of the environment upon them was no longer direct but instead was mediated by their *sentiment intérieur.* Lamarck's thoughts on this particular organic phenomenon are important, for the response of the higher animals to felt needs was a major element in his explanation of the way organic change took place.

Lamarck admitted that he was unable to find a precise expression that conveyed exactly what he meant by *sentiment intérieur.*[107] "Conscience" was not quite right, for that suggested that the mechanism in question operated in conjunction with thought and judgment to produce activity, which was decidedly not what he had in mind.[108] By the phrase *sentiment intérieur* he hoped to identify a "feeling of existence," a "very obscure feeling possessed by all the animals having a nervous system sufficiently developed to give them the faculty to feel."[109] An animal, he supposed, could have this "feeling of existence" without being aware of it.[110]

The *sentiment intérieur* was not, however, merely a feeling of existence, dimly perceived if perceived at all. More significantly, it was a well-integrated system that automatically dictated the animal's responses to stimuli. Lamarck defined the *sentiment intérieur* in 1815 as "a power which, aroused by a felt need, causes the individual to act immediately, in other words, in the same instant as the emotion expressed. If the individual is endowed with the faculties of intelligence it nonetheless acts in this circumstance before any premeditation or operation could arouse its *will.*"[111] The *sentiment intérieur,* he explained in the third volume of his *Histoire naturelle,*

> is not a sensation, it is a very obscure feeling, an infinitely excitable whole composed of intercommunicating separate parts, a whole which any felt need may excite, which [once excited] then acts immediately, and which has the power, in the same instant, to cause the individual to act if necessary.
>
> Thus the *sentiment intérieur* resides in the unity of the organic system of sensations, and all the parts of this system are assembled in a common center [*foyer*]. It is in this center that is produced the disturbance [*émotion*] that the *sentiment* in question can experience, and there also resides its power to induce action. All that is necessary for that to happen is that the *sentiment intérieur* be moved by any need whatsoever, where-

upon it will instantly put in action the parts that must be moved to satisfy the need. This takes place without any of those resolutions that we call *acts of will* being necessary.

Without understanding the nature of the phenomenon, the name *instinct* has been given to this cause that makes animals respond immediately to the needs that arouse them. It has been considered like a flame that enlightens them on the actions to be executed, and it has been noticed that it never misleads them. This involves, however, neither enlightenment nor any need of it, for this cause is uniquely mechanical, and turns out to correspond perfectly, like other such causes, to the effects produced. The action it brings about is never false. The felt need excites the *sentiment intérieur,* this excited feeling includes action, and there is never any error.[112]

It must be stressed that in Lamarck's system of classification the sensitive animals, the animals endowed with a *sentiment intérieur* but not with intelligence in addition, constituted all the higher invertebrates: the insects, the arachnids, the crustaceans, the annelids, the cirripeds, the Acephala or Conchifera, and the mollusks.[113] By the testimony of the above quotation, Lamarck believed that the response of an animal to a felt need did not require any *willing* on the animal's part.

Instincts, in Lamarck's view, were merely the result of habits maintained long enough to have developed an organic basis:

The habit of exercising a certain organ or part of the body in order to satisfy recurrent needs causes the *sentiment intérieur,* when its power is exercised, to give to the subtle fluid it displaces a facility in directing itself. Indeed the subtle fluid is given such a facility in directing itself toward the organ or part where it has so often been employed, and where it has traced open routes for itself, that this habit is changed for the animal into a penchant that soon dominates it, and which then becomes inherent to its nature.[114]

Through the inheritance of acquired characters, what for one generation was habit became for later generations instinct:

As the penchants that animals have acquired through the habits they have been forced to contract have little by little modified their internal organizations, thus rendering the exercise of [these penchants] very easy, the modifications acquired in the organization of each race are then propagated to new individuals through generation. Indeed, it is known that generation transmits to these new individuals the state of organization of the individuals that produced them. It results from this that the same penchants already exist in the new individuals of each race even before they have the chance to exercise them, so that their actions can only be of this one kind.

It is thus that the same habits and penchants are perpetuated from generation to generation in the different individuals of the same races of animals, and that this order of things, in the animals that are merely *sensitive,* should not be expected to offer notable variations, as long as there happens to be no change in the circumstances essential to their way of life, which would be capable of forcing them little by little to change some of their actions.[115]

Lamarck characteristically cited examples from the vertebrates rather than the invertebrates when it came to discussing the influence of particular circumstances on animal form. In the section of additions at the end of the *Philosophie zoologique,* however, he suggested, in broad outline at least, how the taking up of different habitats had led to the formation of the higher classes of invertebrates. He supposed that the worms that lived in water rather than inside a host animal had, in the course of time, become highly diversified. Those that had become used to exposing themselves to the air gave rise to the "amphibious insects" like the midges and the mayflies, and these in turn led to the great many insects that lived entirely out of the water. Some of these insects, forced by circumstances to change their habits, gave rise to the arachnids. Certain of the arachnids developed the habit of frequenting water, and step-by-step they gave rise to the crustaceans. The races of aquatic worms that were never exposed to the air, on the other hand, led to the formation of the annelids, the cirripeds, and the mollusks.[116]

When Lamarck came to discuss the vertebrates, he was much more explicit about the influence of circumstances upon the development of habits and forms. Indeed, the popular conception of Lamarck's evolutionary theory derives primarily from the examples he gave of how changes in habits in the higher animals effected changes in their structures. Lamarck's first examples of how organic change takes place were the following:

The bird attracted by need to the water to find there the prey necessary for its existence, spreads the digits of its feet when it wishes to strike the water and move on the surface. The skin that unites these digits at their base thereby acquires the habit of stretching itself. Thus, with time, the large membranes uniting the digits of ducks, geese, etc. have been formed such as we see them today.

But the bird whose way of life habituates it to perch in trees has necessarily the digits of its feet extended and shaped in another way. Its claws are elongated, sharpened, and curved in a hook to grasp the branches on which it often rests.

In the same way one may perceive that the bird of the shore, which does not at all like to swim, and which however needs to draw near to the water to find its prey, will be continually exposed to sinking in the mud. Wishing to avoid immersing its body in the liquid, [it] acquires the habit of stretching and elongating its legs. The result of this for the generations of these birds that continue to live in this manner is that the individuals will find themselves elevated as on stilts, on long naked legs. . . .[117]

Lamarck's initial choice of examples to illustrate the effects of habit upon form appears to be a curious one, insofar as birds were not his specialty. The significance of the particular examples he gave is that they had been used previously by authors like l'Abbé Pluche to illustrate the Creator's wisdom in designing the different species.[118] The fact that each species seemed especially well suited to the conditions of its existence was well noted in the eighteenth century, and commonly explained within the framework of natural theology. Lamarck chose to turn the traditional understanding of such correlations on its head: it was not form that determined habits, he said, but rather habits that determined form.

In his writings after 1800, Lamarck added other examples of how habits had influenced form. To show that the failure to use an organ causes it to become weaker and eventually disappear, he cited the absence of teeth in the whale and the anteater, the rudimentary character of the eyes of the mole and other animals that live in places where light does not penetrate, the lack of eyes (and head) in the acephalous mollusks (organs rendered useless by the development of the mantle in these creatures), the absence of legs in snakes ("though [these parts] be really in the plan of organization of the animals of their class"), the wingless character of many insects, and, to cite effects that become apparent within the life of a single individual, the shortened intestines of consumers of great quantities of alcohol.[119] To show that the frequent use of an organ strengthens it and augments its faculties, he cited the long necks of shore birds, swans, and geese; the long tongues of the anteater and the woodpecker; the forked tongues of flycatchers, lizards, and

ARDEA *Gularis*

When Lamarck first offered illustrations of the effects of habit upon form, he did not mention his now famous example of the giraffe, but instead spoke only of birds. Included was the shore bird. "Wishing" to prevent its body from getting wet, Lamarck explained, this bird contracted the habit of stretching and elongating its legs. As a result, over the ages, the bird's legs became stilt-like. Contrary to what Lamarck's words suggest, conscious intent on the part of organisms played no significant role in Lamarck's theory of organic change. Pictured here is a shore bird from Senegal that Louis Bosc described in 1792 as a new species. From *Actes de la Société d'Histoire Naturelle de Paris* (1792).

serpents; the shapes of the bodies and feet of various herbivorous mammals; the retractable claws of many carnivores; the large hind-legs and tail of the kangaroo; and, among other examples, the long neck and forelegs of the giraffe:

In regard to habits, it is interesting to observe a product of them in the peculiar form and the height of the giraffe (*camelo-pardalis*). This animal, the largest of the mammals, is known to live in the interior of Africa in places where the earth is nearly always arid and without pasturage, obliging it to browse on the leaves of trees and to continually strive to reach up to them. This habit, maintained for a long time by all the members of the race, has resulted in the forelegs becoming longer than the hind legs and the neck being so lengthened that the giraffe, without standing on its hindlegs, with its head raised reaches a height of six meters (nearly twenty feet.)[120]

The results of an animal's efforts to satisfy its needs, Lamarck presumed, were not limited to the development or deterioration of organs and their faculties: if circumstances required it, the organs of an animal could be displaced. This was evidenced by the placement of both eyes on the upper side of the flatfishes.[121] Furthermore, the activities of an animal could bring into existence wholly new structures. This was the case, Lamarck supposed, with the horns and antlers of ruminants:

The ruminants, being able to use their feet only to support themselves, and having little strength in their jaws . . . , can only fight with blows of the head, by directing toward one another the crown of this part.

In their fits of anger, which are frequent, especially among the males, their *sentiment intérieur* directs . . . the fluids most forcefully toward this part of their head, and forms there a secretion of horny matter in some cases or bony matter mixed with horny matter in others, which give rise to solid protuberances: such is the origin of horns and antlers, with which the majority of these animals are armed.[122]

Lamarck had not come to believe in organic mutability as a result of a consideration of the phenomenon of adaptation, and his theory, unlike Darwin's, was not designed with that phenomenon primarily in mind.[123] His discussions of the close correlations between the structures of living things, their habits, and the conditions of their existence appeared when he set out to explain the existence of anomalies in the general scale of increasing complexity—why it was, for example, that snakes did not have legs even though legs were "really in the plan of organization" of their class. But if adaptation per se was not the problem Lamarck set out to explain, his thoughts about use and disuse and the inheritance of acquired characters did account for certain adaptations in a quite reasonable fashion. The blindness of the mole, for example, was explained readily by the assumption that the eyes of moles had degenerated over many generations as the result of disuse. The disproportion in the size of the hindlegs as compared to the forelegs of the kangaroo was easily accounted for by that animal's method of locomotion. In general, functional adaptations—adaptations corresponding to the way an animal used or failed to use its organs—were amenable to a Lamarckian interpretation.

Lamarck did not attribute a significant role in the evolutionary process to *consciously* purposive responses by organisms to changing environmental circumstances. He himself would have been the first to assert that "wishing" or "willing" was of no consequence in all but the highest forms of animal life and therefore could play no major role in the general process of organic change. He unfortunately opened up his views to misunderstanding and ridicule when he spoke of the wading bird "wishing" to keep its body from getting wet and consequently stretching its legs ("voulant faire en sorte que son corps ne plonge pas dans le liquide, il fera contracter à ses pieds l'habitude de s'étendre et de s'alonger").[124] His view of the process of organic change did not really allow for the initiative on the organism's part that this sentence suggested. An animal responded to a felt need according to the mechanistic functioning of its *sentiment intérieur*.

In the additions to the *Philosophie zoologique* in which he talked about the diversification of the higher invertebrates, Lamarck also described the origin of the different kinds of vertebrates. He explained, among other things, how the cetaceans, the ungulates, and the unguiculates had all originated from the "amphibian mammals":

> Those of the amphibians that conserved the habit of going to the shore divided according to their manner of feeding themselves. Some among them, becoming accustomed to browse on plants, such as the walruses and the sea cows, led little by little to the formation of the ungulate mammals, such as the pachyderms, the ruminants, etc. The others, such as the seals, contracting the habit of nourishing themselves from fish and marine animals, led to the existence of the unguiculate mammals, by the means of races which, in diversifying themselves, became entirely terrestrial.
>
> But those of the aquatic mammals that contracted the habit of never leaving the water, and only coming to breathe at the surface, gave rise probably to the different cetaceans we know.[125]

The derivation of these three groups from the "amphibian mammals" was portrayed in a general table Lamarck offered showing "the origin of the different animals."

To Lamarck's contemporaries, Lamarck's most daring discussion

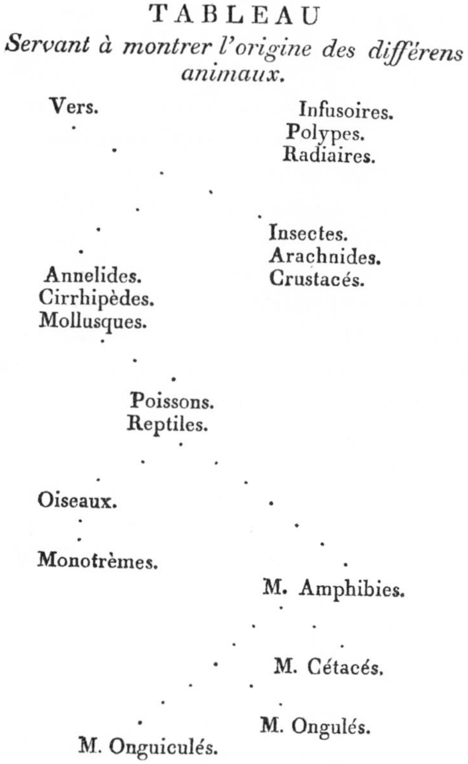

TABLEAU

Servant à montrer l'origine des différens animaux.

Vers.

Infusoires.
Polypes.
Radiaires.

Insectes.
Arachnides.
Crustacés.

Annelides.
Cirrhipèdes.
Mollusques.

Poissons.
Reptiles.

Oiseaux.

Monotrèmes.

M. Amphibies.

M. Cétacés.

M. Ongulés.

M. Onguiculés.

The origin of the different animals, including the derivation of three groups from the "amphibian mammals," as represented by Lamarck in his *Philosophie zoologique* of 1809.

of the influence of circumstances on form was no doubt his discussion of how man himself might have been produced. Lamarck judiciously placed his speculations on this subject within a conditional framework: "If man were distinguished from the animals only by his organization, it would be easy to show that the characters of organization that one uses to form a unique family for man with his varieties are all the product of old changes in his actions and habits that he has taken up and which have become peculiar to the individuals of his species."[126] Lamarck's own views must have been perfectly obvious to his contemporaries despite this initial qualification, for he had made plain elsewhere his belief that the mental as well as the physical phenomena displayed by living things had a wholly organic basis.

Taking a hypothetical quadrumanous race, Lamarck explained how this race would eventually become bimanous if induced to leave its arboreal habitat and to use its feet for purposes of locomotion only. With time, and as the result of responding to various needs, the race would learn to stand and to walk upright. Eventually, as a result of the need to communicate, the race would learn to speak. Lamarck had protested that he considered man "a privileged being, who has in common with animals only that which concerns animal life."[127] He did not specify, however, just what it was that made man privileged. Even the ability to communicate, the faculty Descartes had identified as man's distinguishing feature, was in Lamarck's view readily explicable as an organic modification resulting from a felt need.

Though Lamarck cited a traveler's report on the "orang of Angola" to show how similar one animal was to man, he did not suppose that this creature, "the most perfect of the animals," was actually in the process of becoming man.[128] The existence of man on the globe already, Lamarck shrewdly observed, was a deterrent to the development of similar forms. Explaining himself still in terms of a hypothetical race that was continuing to change its habits, he wrote:

> Hindering the great multiplication of races closely related to it, and keeping them relegated to woods or other uninhabited places, this race will have stopped the progress of the perfecting of the faculties [of the other races], while itself, master to expand itself everywhere, to multiply itself unhindered by others, and to live there in numerous tribes, will have successively created new needs that will have excited its industry and gradually perfected its means and faculties; . . . finally, this preeminent race having acquired an absolute supremacy over all the others, it will succeed in putting between it and the most perfect animals a difference, and, in a way, a considerable distance.[129]

Man, Lamarck believed, was not only the highest form of life that nature had produced—man had made the last stage of the ascent impossible for the forms beneath him.

It was easy enough for Lamarck to explain how an organ could be strengthened or weakened through use or disuse. It was not so easy for him to explain how new organs arose in the first place. He maintained that each class of animals was characterized by a distinct plan of organization, and he was quite specific in identifying what the distinct plan of organization for each class was. His attempts to explain how nature moved from one plan of organization to the next, however, were feeble. He supposed that there were transitional forms linking the adjacent classes in his scheme of classification, and on occasion he tried to specify what some of these forms might be.[130] But all he could offer to explain the appearance of new organs was a "law of organization" that "the production of a new organ in an animal body results from a newly arisen need that continues to make itself felt, and from a new movement that this need gives rise to and maintains."[131] "The forces brought into being by a newly felt need," he insisted, "will necessarily give rise to the proper organ for satisfying this new need, if the organ does not exist already."[132] He admitted that this law was "very difficult to verify by observation."[133] He claimed, however, that the law followed from the general law of the effects of use and disuse. To support his position he offered an explanation of how certain gastropod mollusks had come to have tentacles on their heads:

> I conceive . . . that a gastropod mollusk, which in creeping along experiences the need to touch the bodies lying in front of it, makes efforts to touch these bodies with some of the anterior points of its head, continually sending there masses of nervous fluid as well as other liquids. I conceive, I say, that as the result of these repeated flowings toward the points in question the nerves that terminate at these points will little by little be extended. Now, as in similar circumstances, other fluids of the animal, particularly nourishing fluids, also flow to these same places, it must follow that two or four tentacles will imperceptibly arise and take shape in these circumstances on the points in question. This is no doubt what happened to all the races of gastropods whose needs have made them take up the habit of touching bodies with the parts of their head.[134]

Lamarck's contemporaries and successors were far from convinced that the appearance and development of new organs could be explained in such a way. It was no more than an "arbitrary supposition," said Georges Cuvier, "that desires and efforts can engender organs."[135] In his copy of the second edition of Lamarck's *Histoire naturelle,* Charles Darwin responded to Lamarck's claim by jotting on the page where the claim appeared: "Because use improves an organ, wishing for it, or its use, produces it!!! Oh—"[136]

THE INHERITANCE OF ACQUIRED CHARACTERS

The idea of the inheritance of acquired characters, the idea most commonly associated with Lamarck's name today, was never an issue in Lamarck's mind, and he claimed no originality in espousing it. His basic supposition was straightforward: "Each change acquired in an organ by a habit sufficient to have brought it about is then conserved by generation, if it is common to the individuals that in fecundation join together in the reproduction of their species."[137]

Though Lamarck acknowledged that there were other sources of organic variation besides the development and maintenance of new habits, he did not regard these sources to be of much consequence in the evolutionary process, and by and large he had little to say about them. In the *Philosophie zoologique* he indicated that new varieties could be formed through hybridization and that these would then become races and, through time, "what we call *species*."[138] In the article on species that he wrote for the *Nouveau dictionnaire d'histoire naturelle* he noted that varieties often appear suddenly in plants, but that these varieties can generally be conserved only through grafts or cuttings, and not through sexual reproduction.[139] He observed further that "One must distinguish between varieties obtained accidentally during the development of an embryo, either in a grain, an egg, or a uterus, from those which are formed during the course of the life of the individual; the variety resulting from the first cause being less conservable than that from the second."[140] It did not occur to him that nature's means in arriving at her goals might include or depend upon "accidental" variations.

Maupertuis had considered "varieties obtained accidentally" to be of central importance in the development of organic forms. In Lamarck's own day, Lacépède had identified artificial selection as one of the powerful means man had at his disposal for modifying animals.[141] Lamarck never expressed an appreciation of the roles of random variation and selection in organic change. He attributed the differences in domesticated animals, for example, to the circumstances of their existence—their nourishment, their habits, and the climate in which they lived—rather than to random variation and the choices made by breeders.

Lamarck never attempted to explain the mechanism by which characters acquired by one generation might be transmitted to the following generation. He simply asserted that in sexual reproduction characters acquired by one generation were passed down to the next "provided that these acquired changes are common to the two sexes."[142] He explained his position once by describing a hypothetical experiment:

If one were to take two new born infants of different sexes and mask the left eye of each for life; if then they were united together and the same thing was done with their children, never uniting them except with each other; I do not doubt that at the end of a great number of generations their left eyes would naturally disappear . . . After an enormous amount of time, the necessary circumstances remaining the same, the right eye would come little by little to shift its position.[143]

Significantly, he never attempted to design a feasible experiment to test his views. He presumed that the reality of the inheritance of acquired characters was attested by numerous facts and was not in question.

Lamarck recognized that "in sexual fecundations, mixtures between individuals that have not equally undergone the same modifications in their organization offer something of an exception to [the law that all that has been acquired . . . in the organization of individuals during their lives is conserved by generation]."[144] In such circumstances, he observed, the acquired changes were either not transmitted to the offspring or transmitted only partially: the law of the inheritance of acquired characters in these cases had "only a partial or imperfect application."[145]

Accepting the notion that parents which had not undergone the same modification often had offspring whose characters fell somewhere in between the parental characters, Lamarck admitted that interbreeding could nullify certain "peculiarities of form" that might otherwise form the basis for the development of a distinct race. As he explained this in the *Philosophie zoologique:*

If, when some peculiarities of form or any defects whatever happen to be acquired, two individuals in this situation should unite, they would reproduce the same peculiarities; and with successive generations limiting themselves to similar unions, a peculiar and distinct race would be formed from them. But perpetual mixtures between individuals that do not have the same peculiarities of form would cause all the peculiarities acquired through particular circumstances to disappear. From that one can be sure that if men were not separated by the distances between the places where they live, interbreeding would cause the general characters distinguishing the different nations to disappear.[146]

Lamarck did not, however, regard interbreeding as a serious threat to his theory. This was no doubt because he assumed that a given environment would cause the individuals of a given species to experience the same needs, respond to these needs in the same fashion, and thus undergo the same organic changes. Though as a systematist he recognized the existence of individual variations within a species, as an evolutionary theorist he advanced a typological view of species change: he considered the changes a single individual experienced in responding to the environment to be representative of the changes experienced by the entire race or species to which the individual belonged. Consequently he did not believe that geographical isolation was absolutely necessary for speciation to take place. Furthermore, on the basis of his examination of extensive natural-history collections, he was convinced that species in nature grade into one another.[147]

THE EVIDENCE FOR EVOLUTION

Lamarck had relatively little evidence to support his evolutionary theory. To substantiate his general claim that nature had begun with the simplest forms of life and then successively produced all the others, he could only point to his arrangement of the animal classes in a scale of increasing complexity and then claim that this scale represented the order of formation of the different living things. To substantiate his claim that changing circumstances could, over time, bring about changes in races and species, he had little to offer besides the sort of evidence that had previously been accommodated within explanations of organic diversity that considered organic mutability to be strictly limited. The fossil record was much too incomplete to be of any help to him (and given his assumption that the order of formation of the different living things was represented by the *existing* classes of animals, he had no feeling that the fossil record was crucial for understanding the basic history of life).

"All botanists know," Lamarck wrote in the *Philosophie zoologique,* "that the plants which they transport from their native place to gardens, in order to cultivate them there, undergo changes there, little by little, which finally render them unrecognizable... The effects of changes in circumstances are so well known, that botanists do not like to describe plants in gardens unless they are newly cultivated."[148] One would not be able to find in nature, he indicated, a plant identical to wheat, for man had brought wheat to its present state through cultivation. The same was true of man's cabbages, lettuces, and other vegetables and an "abundance of animals that domesticity has changed or considerably modified": "How many very different races of our chickens and domestic pigeons have we obtained by raising them in different circumstances and in different countries, and which one would search in vain now to find as such in nature!"[149] The differences between the various races of dogs, the marked contrast between the English racehorse and the French workhorse, and similar facts testified, in Lamarck's opinion, to the powerful influence of the environment upon living things.

To the evidence from the domestication of plants and animals that seemed to favor the hypothesis of organic mutability, Lamarck added evidence from comparative anatomy of the sort that seemed to demonstrate the effects of habit upon form. Many of his examples have been cited above. He was aware, for instance, that though adult whales have no teeth at all, rudimentary teeth are to be found in whales in the fetal stage. He did not develop the implications of this discovery at any great length, but he did indicate that the whale's lack of teeth, when most other animals of the same class had teeth, was evidence of an organic change deriving from the adoption of particular habits.[150] The lack of hindlegs in the cetaceans was another example of the dramatic effects habit could have upon form.[151]

Though insisting upon the importance of different circumstances in the production of organic diversity, Lamarck said surprisingly little in his major writings about the geographical variation of living things. His most extensive discussion of the subject appeared in 1817 in the article on "species" that he wrote for the *Nouveau dictionnaire d'histoire naturelle*. There he observed that if one chooses a well-known species from one's own country, and then seeks out the same species when traveling elsewhere, one will in general be able to find the species, but one will begin to notice changes in it as one goes farther from one's point of departure. These changes, Lamarck maintained, will at first be barely perceptible, but will later become so appreciable "that if one compares the last individuals observed with those that one knew first, one will in no way hesitate in regarding them as belonging to distinct species." What is more, Lamarck noted, "one will not find merely a simple series of varieties leading gradually to the distinct species; one will be able to observe that among the varieties obtained there are often those that belong to lateral series, which lead to still other species."[152] Lamarck cited a shell Olivier had found in Egypt that appeared to be a much changed variety of the *Helix pomatia* known in the north of France. The same species had been found in Italy displaying less marked differences from the shell of the north of France, and specimens from the south of France resembled even more closely the type from the north.[153] A similar situation existed, Lamarck maintained, with the cabbage butterfly: "If we follow the cabbage butterfly (*Papilo brassicæ,* L.), so common in our lands, we will observe different varieties of it, and step by step we will see these varieties lead, in other lands, to the races that we characterize as species . . . Each of them, undoubtedly, is constant and always reproduces its like in the circumstances in which it habitually lives; it will not change as long as these circumstances remain the same."[154]

Not many naturalists were in a position, Lamarck acknowledged, to make observations of the sort he had indicated, and thus the kind of geographical variation he was describing had not been fully documented. He believed, however, that what was known on the subject showed his view to be well founded. He valued highly the observations of Péron, the naturalist on the Baudin expedition, who had been "struck with astonishment in comparing successively the objects he collected."[155] Lamarck felt that his own experience with rich natural-history collections had enabled him "to know just how far the species [in certain large genera] blend into one another."[156] Indeed he maintained that only those who had worked at length with such collections could appreciate the extent to which "among the living productions of nature" everything is "more or less *nuancé*."[157]

Lamarck cited the existence of varieties, the difficulty of distinguishing species from one another in large collections, and geographical variation as facts that proved that species did not have an absolute constancy, were not "as old as nature," and had not "all existed originally just as we see them today."[158] In doing so, however, he was only destroying a straw man. The existence of varieties and the difficulty of distinguishing species from one another were well known to his contemporaries. His contemporaries, by and large, did recognize that living things were susceptible of certain modifications. His contemporaries were not willing, though, to take the existence of varieties and the difficulty of distinguishing species as proof of species transmutation. Lamarck appreciated, in some measure at least, the weakness of his argument, and tried to put his opponents on the defensive:

> [To my suggestion that the existence of varieties casts doubt upon the immutability of species] it will be replied that circumstances that have changed and become habitual can, in truth, cause species to vary a little, but without taking them too far from their type, which always stays the same. To this response I will reply . . . that here one explains without furnishing proof, for nothing is presented that positively attests that the type of the species has never changed, and the allegation that those that we see are constant, the circumstances in which they are observed being likewise, does in no way furnish the proof required.[159]

At the same time that Lamarck tried to shift the burden of proof to his critics, he tried to disarm them of an argument that he knew would be used against him. He granted that "assuredly nothing can exist except by the will of the sovereign author of the universe and of nature."[160] He observed, however, that this meant that the Creator was free to bring living things into existence by whatever method He wished. To Lamarck it appeared evident that it was only by studying nature that man could ever hope to learn the method the Creator had actually employed in bringing all things into existence. Lamarck maintained that his own view—that the "supreme author" of everything created matter and an *order of things* which, acting upon matter, successively formed all the bodies we observe—was a view that in no way detracted from the "supreme author's" greatness.[161] Lamarck recognized that his view would put a considerable strain on the imaginations of his contemporaries:

> Would one dare carry the spirit of system so far as to say that it is nature alone that has created this astonishing diversity of means, ruses, cunning, precautions, and patience of which the *industry* of animals offers us so many examples! Is not what we observe simply with the class of *insects* a thousand times greater than what is necessary to make us perceive that the limits of nature's powers in no way allow her to produce so many marvels herself! And to force the most obstinate philosopher to recognize that here the will of the supreme author of all things has been necessary and has alone been sufficient to bring into existence so many admirable things?[162]

Lamarck's answer to this question was again that man should not assign limits to the Creator's power: "If I discover that *nature* herself . . . creates organization, life, even feeling; that she multiplies and diversifies, within bounds that are not known to us, the organs and the faculties of the organized bodies whose existence she sustains or propagates; that she creates in the animals, by the sole means of the *need* that establishes and directs the habits, the source of all the actions, from the most simple up to those that constitute *instinct, industry,* and finally *reasoning;* must I not recognize, in this faculty of nature . . . the execution of the will of her sublime author, who has been able to will that she have this faculty?"[163] There is no evidence that Lamarck was insincere in his references to the "supreme author of all things." Such references could not, however, have been reassuring to those of his contemporaries who believed that the Creator's role in the Creation was a much more direct one.

In explaining how nature had proceeded from the simple to the complex, how plants and animals had changed according to the particular circumstances to which they had been exposed, and how the various faculties had emerged with the development of new systems of organization, Lamarck had in essence followed through on his conclusion of the 1790s that life was not an inconceivable principle. He had put some of the "large facts" relating to living things in an evolutionary framework. It was his great achievement as a naturalist-philosopher. To many of his contemporaries, however, it appeared that he had indeed carried the *esprit de système* too far.

CHAPTER SEVEN

The Frustrations and Consolations of the Naturalist-Philosopher

In concluding his *Philosophie zoologique,* Lamarck observed: "men who strive in their works to push back the limits of human knowledge know well that it is not enough to discover and prove a useful truth that was previously unknown, but that it is necessary also to propagate it and get it recognized."[1] Judged in his own terms, Lamarck as an evolutionary theorist was a failure. Though he was painfully aware of the difficulty of getting novel views recognized, he showed little strategic sense in his attempts to overcome this difficulty. The general arguments that seemed so compelling to him were insufficient to convince his contemporaries of the reality of evolution. But he offered his contemporaries little else.

Few comments on Lamarck's evolutionary views appeared in print in his lifetime. Of these, one of the first and also one of the most instructive came from L. A. G. Bosc, a member of the Société d'Histoire Naturelle de Paris, the Société Philomathique, and the Linnaean Society of London. Bosc, writing in 1802 the section on conchology for the Castel edition of Buffon's *Histoire naturelle,* was enthusiastic about Lamarck's *Système des animaux sans vertèbres,* which had appeared the previous year. Bosc's enthusiasm, though, was directed explicitly at the systematic part of Lamarck's work, not the theoretical comments with which Lamarck opened the book. Bosc lauded Lamarck's willingness to create genera on the basis of single species (this, Bosc explained, produced a conchological system that could accommodate future discoveries in a way that the systems of Linnaeus and Bruguière frequently had not been able to).[2] Bosc also praised the fact that Lamarck described not only shells but the animals that inhabited them as well. All in all, Bosc felt the part of Lamarck's *Système* that dealt with conchology was "certainly the most perfect as a whole that has yet been published."[3] But Bosc was not prepared to follow Lamarck in Lamarck's more philosophical musings. Bosc's treatment of Lamarck's transformist views was brief and unsympathetic:

> Lamarck, at the end of his new work . . . has added an article on fossils in which he maintains that circumstances have brought out a diversity of habits among living things that have led the way to changes in their organization, and that insensibly each acquires a new form, which produces not only new species, but also new genera and orders.
>
> This is no doubt a great idea, and it merits all the attention of the philosopher, but it is not supported by direct proofs, and it is less natural to admit it than to suppose the annihilation of several species, a supposition to which nothing is adverse.[4]

Lamarck's position, of course, was just the opposite of Bosc's. To Lamarck it had seemed much less natural to admit the extinction of species than to admit species mutability. Bosc, however, was more impressed by the disparity between fossil and living forms than by Lamarck's arguments against extinction.

Bosc, without a doubt, was the sort of naturalist Lamarck had in mind when Lamarck spoke of those who concerned themselves only with the identification and classification of nature's productions, only with "small facts" and never with "large facts." Bosc presented more memoirs to the Société d'Histoire Naturelle in its early years than any other naturalist.[5] But what he characteristically offered at the sessions of that society was the identification of some previously unknown species or genus. He was prepared to acknowledge a limited kind or organic mutability. For example, he was willing to suppose that the animals of the univalve mollusk order *Testacella* were the same as slugs, "but modified by climate, the former being found only in hot countries and the others in temperate and even cold countries."[6] He was not prepared, however, to endorse a broad view of organic evolution.

Bosc expressed no concern over the materialistic nature of Lamarck's views. Others expressed this concern, however, as Lamarck developed his system of evolutionary biology at greater length. Writing about the soul, the mind, materialism, fatalism, and free will, the phrenologists F. J. Gall and J. C. Spurzheim criticized Lamarck's views and expressed their own preference for the Biblical account of Creation.[7] The arch-conservative political philosopher Louis de Bonald also condemned Lamarck's views as materialistic.[8] Pierre André-Latreille, the leading entomologist of the day and Lamarck's demonstrator at the museum, was not inclined to speak out against the man who had been like a father to him, but Latreille found it impossible to conceive that the forms and in particular the instincts of insects could be explained by anything other than the work of a wise designer. When he became professor at the museum after Lamarck's death, Latreille presented a creationist view of life to his students and told them: "If we are wrong, do not seek to destroy illusions that are useful rather than harmful to society, and which make us happy or console us in the difficult pilgrimage of life."[9] The leading entomologists in England had similar objections to Lamarck's evolutionary ideas. What William Kirby (a clergyman) and William Spence could not tolerate, they said, was Lamarck's "denying to the Creator the glory of forming those works of creation, the animal and vegetable kingdom . . . , in which His glorious attributes are most conspicuously manifested.[10] The insects, Kirby and Spence felt, were excellent examples of the Creator's handiwork:

> All along, where the uses of any particular organ or part have been ascertained, if you consider its structure with due attention, you will find in it the nicest adaptation of means to an end . . . which proves most triumphantly, that the POWER who immediately gave being to all the animal forms, was neither a blind unconscious power, resulting from a certain order of things, as some philosophists love to speak; nor a formative appetency in the animals themselves, produced by their wants, habits, and local circumstances, and giving birth, in the lapse of ages, to all the animal forms that now people our globe; but a Power altogether distinct from and above nature, and its ALMIGHTY AUTHOR.[11]

Kirby and Spence referred to Lamarck's notions on organic change as "absurd and childish,"[12] but the enormity of Lamarck's theory was evidently such that Kirby could not leave the theory alone, even after Charles Lyell had written an extensive critique of it. Kirby wrote one of the Bridgewater Treatises, entitling his contribution *On the Power, Wisdom, and Goodness of God, as Manifested in the Creation of Animals, and in their History, Habits, and Instincts.* He began his treatise by confronting Lamarck's views directly. Lamarck, Kirby was willing to acknowledge, was a naturalist "distinguished by the variety of his talents and attainments, by the acuteness of his intellect, by the clearness of his conceptions, and remarkable for his intimate acquaintance with his subject."[13] This did not prevent Kirby from feeling he could prove the "utter irrationality" of Lamarck's hypothesis. The facts of adaptation illustrated by the hive-bee seemed adequate to Kirby to demonstrate the strength of his own position. The bee's long tongue, honey stomach, wax pockets, jaw form, and hind-leg structure represented "a number of organs and parts that must have been contemporary, since one is evidently constructed with a view to the other." The whole social system of the hive as well, with the males, females, and workers each performing particular functions, seemed to point to "the end that an intelligent Creator intended." The whole animal kingdom, Kirby maintained, displayed "the same mutual relation and dependence between the different parts and organs and their functions." It therefore seemed to him unthinkable that

> any one in his rational senses [could] believe for a moment that all these adaptations of one organ to another, and of the whole structure to a particular function, resulted originally from the wants of a senseless animal living by absorption, and whose body consisted merely of cellular tissue, which in the lapse of ages, and in an infinity of successive generations, by the motions of its fluids, directed here and there, produced this beautiful and harmonious system of organs all subservient to one purpose.[14]

Kirby had no difficulty pinpointing the primary source of Lamarck's folly:

> Lamarck's great error, and that of many other of his compatriots, is materialism; he seems to have no faith in any thing but *body,* attributing every thing to a physical, and scarcely any thing to a metaphysical cause. Even when, in words, he admits the being of a God, he employs the whole strength of his intellect to prove that he had nothing to do with the works of creation. Thus he excludes the Deity from the government of the world that he has created, putting nature in his place.[15]

Kirby was not just concerned with the way Lamarck's system explained the structures of insects. Kirby was also concerned with all that Lamarck had to say concerning mental faculties, and particularly the mental faculties of man. Lamarck, Kirby explained to his readers, "considers [man's] intellectual powers, not as indicating a spiritual substance derived from heaven, though resident in his body, but merely as the result of his organization, and ascribes to him in the place of a soul, a certain *interior sentiment*."[16] But Lamarck, Kirby said, made the mistake of confounding sensation with intellect, of accepting as real only that which the senses could perceive. There was something that could conduct man farther than his mere sense impressions, and that was *thought*. Said Kirby confidently, "we cannot help *feeling* that our thoughts are the attributes of an immaterial substance." It was thought that enabled man to "take flights beyond the bounds of time and space, and enter into the Holy of Holies." It was thought that could carry man "back to hail with the angelic choirs, the birth-day of nature and of the world which we inhabit." "Who can believe," Kirby asked, "that such a faculty, so divine and god-like and spiritual, can be the mere result of organization?" Matter and spirit, Kirby maintained, were not homogeneous, and no juxtaposition of *material* molecules could ever produce a reasoning being.[17]

Unlike Kirby, most of Lamarck's immediate contemporaries displayed little need to elaborate on Lamarck's errors. Gall and Spurzheim, for example, felt able to respond to Lamarck's theory "in just a few words" since they did "not believe . . . that this bizarre opinion can find many partisans."[18] René Tourlet, the French writer who reviewed Lamarck's biological treatises for the *Moniteur universel,* treated some of Lamarck's biological views at length but seemed to shy away from discussing Lamarck's thoughts on organic mutability. Tourlet, a man of letters with a scientific background, was not hostile to Lamarck's attempt to provide a systematic body of knowledge regarding living things. He praised, for example, Lamarck's idea of beginning with the simplest forms of life to find out what the essence of life really is.[19] But aside from a brief comment to the effect that he doubted that needs could create organs,[20] Tourlet scarcely raised the issue of organic change in his review of Lamarck's *Recherches sur l'organisation des corps vivans,* and though in his review of Lamarck's *Philosophie zoologique* he did mention Lamarck's thoughts on the progressive development of living things, Tourlet's main interest in Lamarck's work seemed to lie in Lamarck's general thoughts on the nature of life.

Lamarck's major biological treatises, it is true, were not solely about organic evolution. They also dealt with the faculties associated with particular structures, the physical causes of life, and the production of mental phenomena. Nonetheless, it seems peculiar that Tourlet should have said so little about Lamarck's evolutionary theory. Tourlet did not hesitate to offer some astute comments on the difficulties Lamarck's books could pose for the reader. The author of the *Recherches sur l'organisation des corps vivans,* Tourlet observed, had not only sewn some seminal ideas but he had also leapt some intellectual precipices, "without noticing the difficulty one may have in following him."[21] No other contemporary of Lamarck captured the character of Lamarck's writings so succinctly.

Perhaps Tourlet did not wish to air the subject of evolution publicly. If that was indeed the case, then he was in effect pursuing the course that was being taken by Lamarck's most effective critic: Georges Cuvier, Professor at the Muséum d'Histoire Naturelle and the Collège de France, member and perpetual secretary of the natural sciences of the first class of the Institut de France, member of the Académie Française, counselor of the Université de France, and presiding officer of the Interior Department of the Council of State (to name only some of his official positions).

In 1795, when he first lectured in Mertrud's place at the museum, Cuvier told the students studying comparative anatomy that he would be satisfied to be a Perugin (the Italian painter who started Raphael on his career).[22] In 1796, when he explained to his distinguished audience at the first public session of the institute what issues were being raised by his study of fossil vertebrates, he presented himself as a modest anatomist who limited himself to the examination of the details under his eyes and scalpel, one who would be content to have opened the subject of the earth's history to the genius who would dare to study it.[23] By 1812, however, when Cuvier suggested that natural history, like astronomy, would one day have its Newton, no one could have doubted that Cuvier considered himself the most eligible candidate for that position, despite his continual insistence that he was not an expounder of bold views but rather a scientist who never went farther than the facts would allow him.[24]

Cuvier's magisterial and disapproving presence has long been recognized as a major factor in the poor reception of Lamarck's evolutionary theory by his contemporaries. Often enough Cuvier's rejection of organic mutability has been seen as a case of scientific thinking being shackled by religious preconceptions. It has been argued recently, however, that the whole of Cuvier's zoological work

Cuvier. The very picture of self-assurance and authority.

rested upon a philosophy of stability, which, independently of any religious considerations, precluded the idea of organic mutability.[25]

Lecturing in 1805 to a fashionable lay audience at the Athenée de Paris, Cuvier allowed that the practitioners of natural history had not been able to agree upon the true object of their science. Some, like Réaumur, had pursued curious facts primarily for the mere enchantment of them. Others, like the professed disciples of Linnaeus, had treated natural history as if it were basically an exercise in nomenclature. Still others, seeking to imitate Buffon, had constructed vain philosophical systems. Only a very few naturalists had followed the example of Aristotle and had attempted to reduce natural history to a body of doctrine.[26] Cuvier was first and foremost a comparative anatomist, and it was in comparative anatomy that he sought the doctrines or rules on which a true science of natural history might be based.

The guiding idea of Cuvier's work as a comparative anatomist was the functional integrity of the living organism. This idea found expression in his two anatomical rules: the "correlation of parts" and the "subordination of characters." It appeared to him that "the laws that determine the relation of organs" were founded upon the functional interdependence of these organs and possessed "a necessity equal to that of metaphysical or mathematical laws, since it is evident that the proper harmony between organs that act on one another is a necessary condition of existence of the being to which they belong."[27] Was it the case, Cuvier asked himself, that nature had employed every conceivable combination of organs or organ systems in her different organic productions? His conclusion was that she had not. Some combinations had not been realized because the organs or organ systems involved were incompatible with each other and thus could not function together as an integral whole. Every living creature had parts and functions that were perfectly correlated one to another.

It was on his observations concerning the compatibilities or incompatibilities between different sorts of animal parts that Cuvier ostensibly based his reconstructions of whole fossil animals from only partial remains. Describing to the members of the first class of the institute his studies of a large fossil quadruped from Paraguay, Cuvier told them that the organs of living things were subordinated to one another in such a manner that those of the first order of importance brought along with their own various conformations the conformations of the other organs of the body. One had only to peruse Jussieu's *Genera Plantarum,* Cuvier said, to learn that an apparently slight peculiarity in the major organs of a plant enabled the botanist to anticipate a great deal of the structure of the rest of the plant. The same relation, Cuvier said, held true of the fossil from Paraguay. Simply on the basis of an acquaintance with its unguiculate toes and its lack of incisors and canines, the zoologist could predict in advance the order to which it belonged. And the remaining parts of the fossil happily confirmed the prediction.[28]

The idea of the functional integrity of the organism did not serve Cuvier only as a basis for reconstructing fossil forms. It was also central to his denial of the reality of significant organic change.

He supposed that no part of an organism could be significantly modified without destroying the harmony between the organs that was necessary for the organism's existence. He recognized the presence of variations within species, but he presumed these to be of a superficial nature that did not denature the species type. The "conditions of existence" of each being, by which Cuvier meant not only the necessary coordination of its parts but also its relations with the beings around it, did not allow variations to proceed very far.[29]

Just how far Cuvier was from embracing the idea of organic mutability can be seen in his belief that the science of comparative anatomy could not exist unless it was founded on a system of stable types. In the 1799 version of his memoir on fossil and living elephants, he broached the possibility that present-day species could have descended from earlier forms as the result of the influence of climatic changes. Having broached the idea, he rejected it:

> Whatever influence climate might have in varying animals, it surely does not go very far; and to say that it can change all the proportions of the bony framework and the fine structure of the teeth, would be to propose that all quadrupeds can be derived from but a single species, that the differences they display are only successive degenerations. In a word, *this would be to reduce to nothing all of natural history, since its object would then consist only of variable forms and fleeting types.*[30]

Beyond his idea that the functional harmonies of the parts of living things, as well as the relations of living things one to another, precluded species from changing significantly, and beyond his belief that there could not even be a science of natural history if organic forms were mutable, Cuvier had additional reasons for opposing the general idea of organic mutability and the specific theory of organic change that Lamarck was propounding. If species had changed gradually, then traces of this gradual change should have been visible in the fossil record. Paleontologists should have discovered forms linking the palaeotherium of the past with comparable species of the present. But no such intermediate forms had been found.[31] If species were not constant, then the degrees of variation they exhibited should not be constrained within narrow bounds. That, however, seemed to be the case. Specific climates or forms of nourishment might produce minor changes in the appearance of members of a species, but they did not make the slightest difference when it came to a character such as the number of bones or their articulations.[32] The production of new forms through the hybridization of pre-existing species was not found in nature, and when man attempted to cross different species he inevitably found that the fecundity of the new line could not be maintained. Even those animals that had been domesticated for innumerable generations varied only in their superficial features.[33]

Some naturalists, Cuvier remarked, counted heavily "on the thousands of centuries that they accumulate with a stroke of the pen," but there was no evidence that time had any more effect on animal characters than climate did. "We can scarcely judge what a long time will produce except by multiplying it by . . . what a lesser time will produce," Cuvier observed, and on the basis of the most ancient remains of modern animals, those embalmed by the ancient Egyptians, there was no reason to believe in anything but the constancy of species. Cats, ibises, birds of prey, dogs, monkeys, crocodiles, and the head of an ox had all been embalmed by the Egyptians, and there was no significant difference between these animals and those of the present day. Though he was only dealing with a time span of two or three thousand years, Cuvier acknowledged, he was going back as far as possible.[34]

As for Lamarck's beliefs in spontaneous generation and a general scale of increasing complexity in the animal kingdom, the first seemed to defy all evidence as well as common sense, and the second failed to recognize that the animal kingdom was represented by a number of basic types that were in essence incomparable and therefore unamenable to a sequential ordering. Cuvier's identification of four distinct plans of animal organization tore apart the chain of being.

There is no question that Cuvier could have assembled a coherent and powerful rebuttal to Lamarck's theory. And yet he did not. This was probably not out of deference to his older colleague, but rather because he never considered Lamarck's theory worthy of his time and effort. Cuvier was not averse to holding Lamarck's thinking up to ridicule. Indeed he did so publicly in 1805 when giving a course on geology at the Athenée de Paris. Marzari Pencati, an Italian naturalist who audited the course, reported the incident to a friend: "The new work on organized beings was ridiculed there as it deserves; the formation of different organs by habits was joked about. And—whatever tenderness the amiable and learned materialist who is the author of it merits—a Cuvier can take the liberty of joking when it comes to animals."[35] Cuvier was disinclined, however, to prepare a detailed rebuttal of Lamarck's thinking for the scientific community. He did not want to dignify Lamarck's speculations with his attention.

Speaking in 1806 on what scientific societies should do to encourage the development of geology, Cuvier observed:

> [Scientific societies] must maintain in regard [to geology] the conduct that they have maintained since their establishment in regard to all the other sciences.
>
> To encourage with their eulogies those who report positive facts, and to retain an absolute silence over the systems that succeed one another.[36]

It would appear that this "absolute silence" with regard to "systems" was the very antidote that had been applied earlier to Lamarck's chemical theory. As Cuvier made evident in his notorious "eulogy" of Lamarck, he considered Lamarck's chemical and zoological theories alike to be "vast edifices [constructed] on imaginary bases," and thus believed it was entirely appropriate for the zoology to receive the same treatment as the chemistry had.[37]

The failure of Lamarck's zoological and evolutionary theorizing, Cuvier maintained in the eulogy, had been evident to all: "Anyone could see for himself that independently of many paralogisms of detail [Lamarck's theory] reposes also on two arbitrary assumptions: first, that it is the seminal vapor that organizes the embryo; second, that desires and efforts can engender organs."[38] Cuvier evidently felt that these assumptions were so preposterous they did not need refutation in the eulogy. He was himself a preformationist, hence Lamarck's thoughts on spontaneous generation and epigenetic development were unacceptable to him.[39] As for the idea that habits could create organs, Cuvier believed that Lamarck had trapped himself in a vicious circle. For a creature to fly, it needed wings. For a bird to swim, it needed web feet. Where could the habits of flying or swimming come from in the first place if these organs had not been present?[40] The fundamental difficulty with Lamarck's theory, Cuvier allowed at one point, was that of demonstrating "why, at the beginning, the reptile [or any other creature] acted against its own nature in adapting habits that were in opposition with its original form."[41] Lamarck's theory, founded on the idea of epigenesis and the idea that habits give rise to organic form, instead of vice versa, was nothing, in Cuvier's mind, that should detain a scientist: "A system resting on similar bases may amuse the imagination of a poet; a metaphysician may derive a whole other generation of systems from it; but it cannot sustain for a moment the examination of whosoever has dissected a hand, a visceral organ, or even a feather."[42]

Lamarck had made the mistake of suggesting that birds had developed the habit of filling their lungs with air in order to make themselves lighter, and that this air penetrated every part of the body, even to the hair follicles, changing the follicles into quills and causing the hair to break up into feathers.[43] This sounded to Cuvier so much like de Maillet's idea that flying fish turned into birds when they fell upon dry land and their scales cracked that he never separated Lamarck's thoughts from de Maillet's: "Whoever dares seriously suggest that a fish, by staying on dry land, would be able to see its scales crack and change into feathers, and itself become a bird; or that a quadruped by penetrating narrow passageways and wiggling along could change itself into a snake, does nothing other than prove the most profound ignorance of anatomy."[44] As Cuvier contemplated the "admirable and complicated organization of the feather"—"its coats, its ducts, its transitory cups that are fit perfectly by its barbs, a part of which remains in its shaft, its barbules of various kinds, so well adapted to the nature of the bird"—he was appalled by the absurdity of the transmutationists' idea that a feather could come from a cracked scale.[45] Cuvier, the great comparative anatomist, felt secure in his skills as a dissector, which he knew Lamarck did not have. The scalpel, he was confident, was the real tool for the naturalist who cared about facts. Shortly after Lamarck's death, when challenges from Étienne Geoffroy Saint-Hilaire and the German nature-philosophers finally forced Cuvier to write a direct refutation of the "pantheistic" ideas of the unity of composition and the transmutation of species, Cuvier observed acidly that metaphysics could not change the facts of anatomy: had an unbiased Spinoza ever done any dissections, he would have seen the same bones, muscles, and nerves as did a Boerhaave or a Von Haller.[46]

In his eulogy of Lamarck Cuvier observed: "Whatever interest [Lamarck's zoological works] may have excited by their positive parts, no one believed their systematic part dangerous enough to merit being attacked; it was left in the same peace as the chemical theory.[47] Cuvier's choice of the words "dangerous enough to merit being attacked" rather than some equivalent of "not reasonable enough to merit being considered" suggests that Cuvier did perceive Lamarck's theory as a threat, if only a minor one. His claim that

Lamarck's zoological speculations were "left in the same peace as the chemical theory" also deserves a comment. Lamarck's complaints about the scientific community's suppression of his chemical views seem to apply equally well to its treatment of his biological views. The only peace in which Lamarck's biological theorizing was left was that *public* silence Cuvier recommended for all "systems." Privately, things were different. As Étienne Geoffroy Saint-Hilaire described them, the last years of Lamarck's life were not years of peaceful neglect. Lamarck was "attacked on all sides" with "odious jests" and "cutting epigrams." Lamarck, said Geoffroy, was "too indignant to respond" to his attackers and submitted to these insults "with a sorrowful patience."[48]

Cuvier was one of those who regarded Lamarck's views as a source of entertainment. In a work published posthumously, his comment on authors who had favored the idea of species transformation was that "from the moment these authors wished to enter into detail they fell into ridicule."[49] Frédéric Cuvier said that his brother put ideas of species transformation "in the rank of those frivolous games of the imagination with which the truth has nothing in common; with which one may amuse oneself when they are skillfully and gracefully presented, but which lose all their charm when taken seriously."[50]

Owing to Cuvier's program of public silence on "systems," only a few examples remain of the *plaisanteries* he conceived regarding Lamarck's evolutionary views.[51] The most striking of these is to be found in the manuscript version of the first edition of his *Recherches sur les ossemens fossiles.* In the published version of this work (1812), the introductory discourse contains a lengthy paragraph presenting ideas that are attributed indiscriminately to de Maillet, an obscure German physician named Rodig, and Lamarck.[52] The paragraph, which follows a brief discussion of the theories of the earth of Leibniz, de Maillet, and Buffon, begins:

> In our times, some freer minds than ever have also wanted to exercise themselves on [the subject of the origin of the earth]. Some writers have reproduced and prodigiously extended the ideas of de Maillet. They say that all was fluid in the origin; that the fluid engendered at first some very simple animals such as the monads or other infusorial and microscopical species; that, through time and in taking up diverse habits, the races of these animals became more complex and diversified themselves to the point where we see them today.[53]

The published work continues with a reference to ideas that can be found in Lamarck's *Hydrogéologie*. In the manuscript, however, there appears a caricature that never appeared in print:

> that the habit of chewing, for example, resulted at the end of a few centuries in giving them teeth; that the habit of walking gave them legs; ducks by dint of diving became pikes; pikes by dint of happening upon dry land changed into ducks; hens searching for their food at the water's edge, and striving not to get their thighs wet, succeeded so well in elongating their legs that they became herons or storks. Thus took form by degrees those hundred thousand diverse races, the classification of which so cruelly embarrasses the unfortunate race that habit has changed into naturalists.[54]

The mention of hens changed by habit into herons or storks, exaggerated though it was, did have a counterpart in Lamarck's writings, but the mention of pikes becoming ducks or ducks becoming pikes did not. De Maillet and Rodig had described how a fish might be transformed into a bird. Lamarck, on the other hand, convinced that nature only proceeded step by step, presumed that it was only over the course of countless generations that the fish had given rise to the reptiles and one branch of the reptiles had given rise to the birds. Cuvier did not bother to distinguish between the views of Lamarck and the views of de Maillet and Rodig. It was characteristic of him to lump Lamarck together with the likes of de Maillet, Rodig, and Robinet.[55] Significantly, though, Cuvier refrained from attacking bigger targets. He never acknowledged that Buffon and Lacépède had also endorsed the notion of organic mutability. Buffon's reputation still carried a great deal of weight, and Lacépède, though not truly of the first rank as a naturalist, was a man of considerable stature, being one of the senators of the consulate and then grand chancellor of the Legion of Honor and a minister of state.[56] When in 1800 the second volume of Lacépède's *Histoire naturelle des poissons* appeared—the volume in which Lacépède advanced his own thoughts on species mutability—Cuvier sent Lacépède a letter praising the book and reported favorably on the book to the first class of the institute.[57]

Cuvier's political skills did not fail to attract the attention of his contemporaries. Marzari Pencati recounted that Cuvier had surprised him in the course on geology at the Athenée by endorsing a view of the earth's recent history that was reconcilable with the Bible. It crossed the Italian naturalist's mind, he said, that it was not entirely coincidental that Cuvier's expression of these views followed closely upon the Pope's arrival in Paris: "I thought at first that it was only a cardinal's cap that the good devotee awaited; this conjecture was not extravagant in a city where all are flatterers, and in a time when everyone changes position. If a zoologist of the Muséum was elected Grand Chancellor of the Legion, why should not another of them be made cardinal?"[58] Dropping his sarcasm, however, Cuvier's auditor admitted that he had been unable to detect any sophistry in Cuvier's scientific arguments, and that he had to conclude that Cuvier's position was based principally on his paleontological researches and was sincere enough. Cuvier, it seems fair to say, was not a scientist who let the Bible dictate what was to be considered scientifically acceptable. It is clear, on the other hand, that Cuvier had a keen sense of where the power lay in the scientific community and in society at large, and he did not fail to profit from this in advancing his career.

Cuvier may not have had a better argument than Lamarck on all points. The existence of vestigial organs, such as the teeth found by Geoffroy in the fetus of a right whale or the rudimentary eyes in the *Aspalax* (a creature that burrowed underground) that Olivier discovered in Persia, were in Lamarck's view an indication of ancestral structures that had degenerated because they had ceased to be used.[59] Cuvier admitted the existence of vestiges, but he did not seek to explain them, calling them simply "one of the remarkable peculiarities of natural history."[60] In Cuvier's and Lamarck's day, however, the subject of vestigial organs did not become a focus of debate. Too much of the evidence seemed to be on Cuvier's side, and Lamarck, thanks especially to his use of some ill-considered examples, had left himself open to ridicule.

Even Étienne Geoffroy Saint-Hilaire, who was sympathetic to Lamarck and came to advance transformist views of his own, had to admit that Lamarck's presentation of his ideas suffered from some "great flaws in execution." "In order to arrive at the demonstration of the true principle of the variability of forms in organized beings," Geoffroy explained, "Lamarck too often produced profuse, exaggerated, and for the most part erroneous proofs, which his adversaries, adept at seizing the weak side of his talent, hastened to pick up and bring to light."[61] Lamarck himself may have come to realize that the examples he had offered in support of his evolutionary views had done him more harm than good. He had, at least, come to a similar realization in an analogous situation: his offering of meteorological probabilities in conjunction with his meteorological theorizing. Looking back in his eleventh and final *Annuaire météorologique* (1810) on the *Annuaires* he had published for more than a decade, he admitted a strategic error on his part in not treating the probabilities he had presented seriously enough: "I perhaps greatly wronged the study that I wanted to encourage, supposing incorrectly that more attention would be paid to the observations recorded in the different numbers of the *Annuaires* than to the probabilities presented there."[62] In his evolutionary theorizing, Lamarck undoubtedly felt that his examples of how evolutionary change took place were inconsequential in comparison to the broad principles he wished to make known. His examples of change, however, were what caught the attention of his contemporaries and successors the most. His theory was visualized, caricatured, and ridiculed in large measure on the basis of the examples he offered in its support. In his last major presentation of his theory, in the introduction to his *Histoire naturelle des animaux sans vertèbres,* he omitted all his previous examples of the effects of habit upon structure.[63] The damage, however, was already done. Today, if Lamarck's name summons up any image at all, it is inevitably the image of the giraffe stretching its neck to reach the leaves above it.

Precisely how much Cuvier's treatment of Lamarck's evolutionary ideas influenced contemporary naturalists is impossible to determine, but there is no question that it was a weighty example for a vast majority of the naturalists of the day.[64] Mocked in private and stifled in public, Lamarck's views never took hold in the French scientific community. As Raspail, the sharp-tongued observer of the French scientific scene, remarked in the year after Lamarck's death: "among us ridicule is a deadly weapon; all its blows are mortal."[65]

IT IS OFTEN SAID that Lamarck's evolutionary theory was rejected in its own day simply because people at the beginning of the nineteenth century were unaccustomed to thinking in evolutionary terms. Lamarck, in other words, was too far ahead of his time to be appreciated. What seems to be more nearly the truth, at least with respect to the French scientific community, is that Lamarck's theory of evolution was rejected not because the idea of organic mutability was virtually unthinkable at the time, but because Lamarck's support of that idea was unconvincing and because, more generally, the kind of speculative venture Lamarck had embarked upon did not correspond with contemporary views of the kind of work a naturalist should be doing.

In France at the beginning of the nineteenth century, Lamarck was by no means the only naturalist toying with the idea of organic mutability. Indeed, of the zoologists at the Muséum d'Histoire Naturelle in the early 1800s, only Cuvier was clearly unsympathetic with the idea of organic mutability. Cuvier occupied the chair of comparative anatomy, first as a substitute for Mertrud and then, as of 1802, in his own right. The three chairs devoted specifically to zoology were occupied by Lamarck (professor of the insects, worms and microscopic animals), Lacépède (professor of the reptiles and fish), and Geoffroy Saint-Hilaire (professor of the mammals and birds). Lacépède as well as Lamarck began advancing ideas on organic mutability in 1800. Geoffroy Saint-Hilaire may have entertained ideas on organic mutability as early as the 1790s, though he did not clearly set out transformist views of his own until the 1820s.[66] Faujas de Saint-Fond, the professor of geology at the museum, also expressed interest in the idea of organic mutability, and Cuvier regarded Faujas as a believer in species change.[67] It is thus apparent that Lamarck's belief in organic mutability was not as anomalous as has commonly been supposed. It remains true, however, that Lacépède, Geoffroy Saint-Hilaire, and Faujas de Saint-Fond had differing opinions regarding the history of life on the earth, and that none of these opinions were identical to Lamarck's.

Faujas wrote in 1799 of the possibility that nature had proceeded progressively from the simple to the complex in bringing living things into existence.[68] This was not, however, an idea that Faujas developed further in his later writings, and from what he did publish it is evident that he did not go along with Lamarck's broad, uniformitarian views.[69] Faujas was prepared to admit both the reality of species extinction and the occurrence of major upheavals in the earth's past.

Lacépède was much more explicit than Faujas on the subject of organic mutability. He introduced the second volume of his *Histoire naturelle des poissons* (1800) with a discourse entitled "On the duration of species."[70] As Lamarck would do two years later, Lacépède set forth a view of species that was basically nominalistic: "Why not proclaim an important truth? The species is like the genus, the order, and the class: it is basically an abstraction of the mind, a collective idea, necessary for apprehending, comparing, knowing, instructing. Nature has only created beings that resemble each other and beings that differ."[71]

Acknowledging that the naturalist could not do without the species as the conventional unit of his researches, Lacépède identified two ways in which a species could disappear: either through a violent catastrophe or through "a long train of imperceptible nuances and successive alterations."[72] Catastrophes, Lacépède supposed, were "blows that Nature strikes only rarely."[73] Species could also die out, however, through a general enfeeblement of their organs or—quite the opposite—through an increase in vitality of such magnitude that the individuals of the species would in effect be broken apart.[74] In all of these cases the species would disappear without leaving any descendants. But there was in Lacépède's view another means by which a species could disappear:

> The species can undergo such a large number of modifications in its forms and qualities, that without losing its vital capacity, it may be, by its latest conformation and properties, farther removed from its original state than from a different species: it is in that case metamorphosed into a new species. The elements of which it is composed in its later form are of the same nature as formerly, but their combination has changed: it is truly a second species which succeeds to the old one.[75]

Lacépède offered an arbitrary classification of the different ways in which a species could change—in consistency and nature of parts, in size, in shape, in exterior characters, in interior characters, in sensibility—but he did not describe the mechanisms that would produce these changes. He did mention that species were changed in different ways by nature and by art, noting in particular that man had a special power at his disposal in the form of selective breeding.[76] He did not believe that selection played a significant role in species change in nature.[77] The natural mechanism he seems to have had most in mind was that familiar eighteenth-century offering, the influence of the environment. Fishes, he suggested, would be gradually modified as they found themselves in waters that were muddier or clearer, slower or swifter, hotter or colder than the waters to which they were accustomed.[78] Like Lamarck, Lacépède supposed that any given place on the earth's surface would be subject to climatic change in the course of time, and the species there would have to change accordingly, migrate, or die out.[79] Lacépède did note some phenomena to which Lamarck never really addressed himself, such as the persistence of forms under changing conditions and the variation among the offspring of a single mating.[80] But Lacépède did not deal with these at any length.

Lacépède seems to have been less concerned with treating his subject critically than with presenting a rhapsodic view of nature's wonders. He admitted that some species of fish had probably become extinct. But, he said, considering how much of the earth and seas was as yet unexplored by man, "it is only with a great reserve that we ought to say that a species has terminated its duration."[81] "How many rivers, lakes and places are unknown! And of these habitations as yet hidden from our researches, how many contain species more or less analagous to those of the living individuals or fossil remains we have described!"[82]

Lacépède was willing to leave the subject of his inquiry to future investigators, to a time when science would be more advanced, when the whole world would have been explored, and when many more living and fossil forms would have been compared. Then, he said, when man's genius

> interrogates Nature in the name of Time, and Time in the name of Nature, what fruitful comparisons will arise in all respects! What admirable results! What sublime truths! What immense tableaux! What new light will be shed on the primitive state of the species, on the relations that connected them in ages far removed from our own, on their smaller number in this ancient time, on their more similar sizes, on their more different traits, on their more dissimilar habits, on their more difficult unions, on their longer durations![83]

Analysis fell second in Lacépède's writings to the demands of eloquence. What Lacépède had in mind concerning the origin of life on earth is not clear in the quotation just cited. He was somewhat more specific in a discourse of 1802 on "the effects of human artifice on the nature of fish."[84] He underlined the broad significance of his topic clearly: "The species that man produces, either by his influence upon the individuals submitted to his rule or by the unions he establishes between neighboring or distant species, will be an excellent means of comparison for judging those that nature has given or could give rise to in the course of centuries . . . Artificial species will be the measure of natural ones."[85]

Lacépède (1756–1825), whose thoughts on organic mutability were lost in his raptures over the grand spectacles of nature.

Lacépède did not dwell upon this suggested comparison between artificial and natural species, but he identified four "powerful" means man has at his disposal for modifying animals: (1) the abundance and kind of food he gives to them; (2) the shelter he provides them; (3) the constraints he imposes upon them; and (4) the choice he makes of males and females to propagate the species.[86] He also identified time—"this use of every instant in an uninterrupted period of thousands of centuries"—as the major advantage nature had over man.[87] Concerning the problem of the origin of life on earth, Lacépède envisaged "two major ways of considering the animate universe worthy of all the attention of the true naturalist":

On the one hand, one can picture in very ancient times all the animals existing still in a few primitive species, which by means analogous to those that human artifice can employ, have produced, by the force of nature, secondary species, which by themselves, or by their union with the primitive species, have given birth to tertiary species, etc.... On the other hand, one can suppose that at first every manner of being was used by nature; that she realized all the forms, developed all the organs, put in play all the faculties, gave birth to all the living beings that the most bizarre imagination can conceive; that in this infinite number of species, those that had received only imperfect means of providing for their nourishment, conservation, and reproduction have successively been annihilated; and that everything has finally been reduced to those major species, those better fated beings, which are still represented on the globe.[88]

Lacépède did not declare himself on the matter, but he seems to have been inclined to the first of these views: the development of the many species from the few. But this, it must be pointed out, was quite different from the view that Lamarck was embracing at this time and also quite different from the process of evolution as now conceived. Lacépède's view was essentially Buffon's view, or a combination of Buffon's view with Linnaeus'. It was not the idea that from the simplest forms of life all the others have successively developed. It was instead the idea that from a relatively *few* species all the others have developed, as, for example, all the quadrupeds from a limited number of original quadruped types. The process of organic change, as Lacépède evidently saw it, was one in which the intervals between the original forms were gradually filled up through degeneration and hybridization. Near the end of his life, aware of the gravity of the illness that was to prove fatal to him, Lacépède is reported to have said to his physician: "I am going to rejoin Buffon."[89] The sentiment seems appropriate, for at least in intellectual matters Lacépède never strayed far from the views of his mentor. Lacépède's scientific views bear more evidence of discipleship than originality.

Geoffroy Saint-Hilaire, in contrast to Lacépède, displayed considerable depth and originality in his thinking.[90] The major concern of his later researches and his own battle with Cuvier was not the issue of organic mutability but rather the concept of the unity of composition or type, the idea that all the various animals manifest different degrees of realization of a single, ideal form. Geoffroy too, however, came to believe in organic mutability, and in doing so he found reason to hail the insights of the elderly professor of invertebrate zoology at the museum who had found so little support among his contemporaries.

Geoffroy's thoughts on organic mutability emerged in the 1820s from his work in anatomy, paleontology, and experimental embryology. Beginning in 1825, in a memoir on the anatomy of crocodiles, he suggested the possibility that the crocodiles of modern times had descended from antediluvian creatures, the Teleosaurus and Stenosaurus, even though the differences between the fossil and modern forms were "great enough to have been arranged according to our rules in the class of generic distinctions."[91] In doing so he cited Lamarck's thoughts on the influence of the environment on the organization of living things.

Geoffroy did not adopt Lamarck's ideas on the mechanism of organic change at the species level, however. Though in a report of 1828 he called attention to proof of "the transmission through generation of certain acquired habits,"[92] his own idea was that major changes in the physical conditions of the globe—changes in the temperature and the composition of the atmosphere—had direct effects, teratogenic or "pathological," upon living beings.[93] These changes acted most forcefully upon embryos rather than already formed adults. In the cases where the effects were not fatal, the result was the production of significant organic variation.

Geoffroy's most precise expression of his reflections on organic change were delivered in a memoir to the Académie des Sciences in 1831.[94] The time had come, he indicated, for zoology to move into a new epoch. Identifying the distinguishing characteristics of beings, making an inventory of nature's productions, studying the natural relations of these productions, and proclaiming the unity of organic compositions as the "fundamental and universal fact of organization" were all stages of zoology's past. In the future, Geoffroy hoped, zoology would be perfected in a new way. Geoffroy looked forward to a "philosophical system of differences" in which each organic form would be explained not just by the arrangement of its parts but also by "a second principle of causality; . . . the action of the ambient world."[95] Two different kinds of features, Geoffroy maintained, needed to be considered in the study of organization: "(1) those that belong to the essence of germs, and (2) those that arise from the intervention of the external world."[96] The environment could vary—"cold gives way to heat, the light gases of the atmosphere are replaced by the heavier"[97]—and an embryo that developed one way under one set of conditions might develop a different way under a different set of conditions. Were the environmental changes to last for centuries instead of just a few years, the alterations of organic form would be more profound and more stable.[98]

Geoffroy attached special importance to the function of respiration in the production of new forms. A small "accident" in the development of a reptile, he suggested, could have an "incalculable importance as to its effects" if that accident happened to be a constriction in the pulmonary sac that left all the sanguinary vessels in the thorax and the bottom of the pulmonary sac in the abdomen. Indeed, this accident would "favor the development of the whole organization of a bird; for the air in the abdominal cells would be forced back by the muscles of the lower belly in such a manner as to direct to the respiratory vessels air that was compressed and of the quality of that which leaves a bellows, in other words, air with more oxygen in a smaller volume and consequently with greater energy during the period of combustion."[99] From this first change would come others, including a greater heat, brighter color, and speedier circulation of the blood, a more energetic muscular action, and "the change of the tegumentary tufts into feathers."[100]

In hazarding the views that he did, Geoffroy was not so much offering a completed system as appealing for further study. He distinguished between two theories on the development of organs: "the one supposes the pre-existence of germs and their indefinite encapsulation one within another [*emboîtement*]; the other recognizes their successive formation and their evolution in the course of ages."[101] His own view was the second, the idea of epigenesis. This view, he hastened to point out, had decided advantages over the first, that of the preformationists. The preformationists, he maintained, abridged the field of scientific inquiry unnecessarily. They dispensed with the "study of all the relations that arise from the continual variation of living things" and merely allowed for the description and classification of living things according to their differences. They neglected the means of study open to the ardent scientist and surrendered themselves "to the feelings of the theologian." In accepting the doctrine of the pre-existence of germs, they accepted "a mystery that requires of them no other mental effort than believing it." The epigenesists, on the other hand, had a vast field of enquiry open to their investigations. Geoffroy noted with pleasure that every day the preformationist camp was losing disciples while the believers in epigenesis were gaining in numbers.[102] While preformationists were disposed to reject the idea of organic mutability, however, epigenesists did not by any means automatically endorse it.

Given the views on organic mutability that naturalists other than Lamarck offered in the early years of the nineteenth century, the striking thing about Lamarck is not that he was thinking about organic mutability at a time when no one else was, but rather that at the beginning of the nineteenth century he was the only one who chose to go ahead and develop at length a broad hypothesis of organic change. Lacépède ceased to be active as a naturalist shortly after the appearance of his two discourses touching on species change. Saddened by the death of his wife, and finding himself more and more involved in politics, Lacépède in large measure gave up his scientific career in 1802 or 1803.[103] Geoffroy Saint-Hilaire, whatever his thoughts on organic mutability may have been in the 1790s, or were to be in the 1820s, was not so intellectually adventurous in the first decade and a half of the nineteenth century. Most of his publications in that period were strictly descriptive. Faujas de Saint-Fond, though not averse to theory, was attempting to stress the factual basis of his geological views at the beginning of the new century. More than twenty years earlier, he explained in 1802, he, together with Dolomieu and Saussure, had provided proofs of a major deluge in the earth's history, a deluge that had destroyed many organisms and spread their remains far and wide. He, Dolomieu, and Saussure had not attempted to explain this great accident, Faujas said, for the following reason: "We were unanimously in agreement that before being able to embrace a theory profitably, it was absolutely indispensable to allow the mass of facts to grow, to discuss them, to analyze them one by one, so to speak, before accepting them, and that these facts ought then to be put in reserve as so many choice materials destined for the broad edifice of a theory, facts that would naturally come to arrange themselves in their destined place."[104] This procedure, Faujas said, was no doubt one the impatient would not find agreeable, but it seemed to him to be the best. It was perhaps with such strictures in mind that Faujas refrained from pursuing in print the broad view of organic development he hinted at in 1799. In sum, whatever thoughts on organic mutability Lacépède, Geoffroy Saint-Hilaire, and Faujas de Saint-Fond may have been entertaining at the turn of the century, none of these naturalists conceived the elaboration of a theory of organic mutability as his immediate task. Lamarck felt quite differently. When the idea of organic evolution came to him, he embraced it and made it his own.

UNLIKE Faujas de Saint-Fond, Lamarck was not one to speak of facts "arranging themselves" in their appropriate place. The task of anyone who wished to advance the study of nature, Lamarck indicated in 1802, was "to gather the observed facts and employ them in discovering unknown truths."[105] Well-known facts, he believed, could be united in novel ways. At a time when others were calling for new information, he was prepared to synthesize. Worried by the weakening of his heath, skeptical of the motives of contemporaries who had rejected or ignored his physico-chemical views, and scorning his contemporaries' preoccupation with "little facts," Lamarck pressed forward.

Shortly after coming to believe in evolution, Lamarck entertained the idea of writing a major work on biology that would begin with an exposition of facts rather than general principles.[106] His preference, he acknowledged at the time, was for a deductive style of presentation, but he felt an empirically oriented presentation was necessary to persuade his contemporaries of the validity of his views. His *Biologie* was never executed, however, and its proposed format was never realized in any of Lamarck's other writings. His arguments for evolution took the form of broad assertions accompanied by a limited number of examples but no solid empirical foundation. He played the role of the naturalist-philosopher occupying himself with the "large facts." Most of his contemporaries, conceiving their own tasks differently at just that time, were inclined to look upon Lamarck's evolutionary theorizing as an enterprise that was essentially unscientific.

Lamarck is sometimes cited as one of the founders of the idea of "positive" knowledge. His last work, entitled *Système analytique des connaissances positives de l'homme,* was devoted to outlining what man did know or was capable of knowing positively.[107] Despite what he said about basing his works on facts, however, he proceeded to erect theoretical systems on only the sketchiest of empirical foundations. His statements concerning scientific method—like those of many figures in the history of science—cannot be assumed to be an accurate representation of his scientific behavior.

It is not clear what Lamarck could have done in the first decade and a half of the nineteenth century to have made his theory of evolution more successful. His conception of the evolutionary process included several basic assumptions that modern science has not borne out. What is more, critical evidence supporting the idea of organic mutability was not readily available in the early 1800s. Gathering circumstantial evidence to support the idea of the inheritance of acquired characters would not have been difficult—both Cuvier and Geoffroy Saint-Hilaire, for example, cited evidence in support of the idea[108]—but the inheritance of acquired characters was not the question at issue. What was of concern to naturalists early in the nineteenth century were the transitional forms that Lamarck predicted would be found connecting the major groups of animals. By and large, these transitional forms did not turn up as Lamarck predicted they would. Indeed, Cuvier's idea that the animal kingdom was represented by four distinct plans of organization represented a denial that such transitional forms would ever be discovered. And what failed to be found among modern forms also failed to show up in the fossil record, Geoffroy Saint-Hilaire's claims about crocodiles notwithstanding. The fossil record did not reveal links between the modern classes, nor did it document the transmutation of species.

As fossil evidence continued to come in, what it showed was that Lamarck had been wrong in denying the reality of species extinction, not that he was correct in advocating the idea of organic evolution. In the first several decades of the nineteenth century the fossil record did not even provide conclusive support for the idea that organic forms had become increasingly complex over time. Charles Lyell denied for many years that the fossil record provided evidence for biological progression. So too did the eminent conchologist G. P. Deshayes, who co-edited the second edition of Lamarck's *Histoire naturelle des animaux sans vertèbres* and displayed a certain sympathy with Lamarck's views.[109] It was only some years after the publication of the *Origin of Species* that the fossil record began to become more of a help than a hindrance to evolutionists.

Perhaps Lamarck could have done more with evidence from biogeography than he did. Indeed, his inattention to the facts of biogeography seems surprising when one considers Buffon's writings on the subject, the great interest in *géographie physique* displayed by some of Lamarck's contemporaries, and the importance Lamarck attributed to environmental conditions in his own theorizing.[110]

Lamarck did cite the studies of his friend Olivier and the naturalist-voyager Péron in support of his thoughts on geographical variation, but biogeography never provided him with an argument for evolution the way it did for Darwin and Wallace.[111] Had Lamarck been more a naturalist of the field and less a naturalist of the *cabinet,* he might have thought more about the value of biogeographical evidence.

Lamarck's problem was not limited to a lack of evidence in support of the idea of organic change. He was also unable to provide a satisfactory mechanism to account for the facts of adaptation. While he called attention to the close correlation between organic forms and their special requirements as living things, his thoughts on the creative powers of circulating fluids and on the long-term effects of the use or disuse of organs were inadequate to explain many of the more striking of these correlations. Natural selection, the key to explaining adaptation that was discovered by Darwin and Wallace, never occurred to Lamarck. The balance of nature, as Lamarck saw it, was characterized by "wise precautions" through which the continued existence of each species was assured. Within such a framework neither the idea of extinction nor the idea of change through natural selection was likely to be conceived.

This is not to suggest that scientists were willing to admit the all-sufficiency of natural selection as an evolutionary mechanism as soon as Darwin had proposed it. In fact, perceived inadequacies in Darwin's theory provided the occasion for the so-called "neo-Lamarckian" explanations of organic change that appeared in the last decades of the nineteenth century. But the explanations that bore Lamarck's name only partly resembled Lamarck's own theory. Furthermore, they did not represent a school of thought that had existed, if only clandestinely, since Lamarck's time. No line of discipleship links the neo-Lamarckians of the late nineteenth century with Lamarck himself. Lamarck's contributions as an evolutionary theorist had not been entirely forgotten prior to the publication of the *Origin*. For the most part, however, they were mentioned only to be disparaged along with other heresies, such as the view of organic change that was espoused in the anonymous *Vestiges of the Natural History of Creation*.[112] It was the work of Darwin, not Lamarck, that brought the scientific world to a general acceptance of the idea of evolution. Darwin amassed a great deal of evidence in support of his views, he had a mechanism that accounted for adaptive change more successfully than Lamarck's notions of use and disuse, and he was a shrewd strategist both in the presentation of his views to the public and his cultivation of support within the scientific community.

Lamarck was not pleased by the poor reception that his evolutionary views received. But he had not expected much from the scientific community that had already ignored his physico-chemical views and was misrepresenting his work in meteorology.[113] How it annoyed him to hear his opponents say that they reasoned only according to the facts! He had little patience with those who did not realize that a theory could be founded upon a great number of facts and still be false, and he especially scorned those who concentrated their attentions on what he called the "small" instead of the "large" facts.[114] It seemed incredible to him that those who had done a few experiments in flasks in a laboratory should consider themselves capable of pronouncing on the true laws of nature when they had neglected to contemplate the large facts presented by nature as a whole: facts relative to the production of meteorological phenomena, the organization of living things, and the fate of the remains of living things on the earth's surface. It was as if, Lamarck complained, one tried to build up the science of geography not by first getting an overview of the oceans, continents, islands, mountain chains, rivers, and the like but instead by starting in one spot and slowly moving across the ground with a microscope.[115] But the scientific community, as Lamarck saw it, was controlled by two types of individuals working jointly: a certain number of skillful elocutionists who knew how to make their views prevail and a larger number of persons without real talent who simply added supporting materials to the theories already in vogue.[116] Against such opposition he felt he stood little chance.

As not only a long-time member of the French scientific community but also a witness of the French political scene in the years of the Revolution and afterward, Lamarck was convinced that man was guided less by reason than by narrow self-interest. When, in the second decade of the nineteenth century, Lamarck composed an article on "man" for the *Nouveau dictionnaire d'histoire naturelle,* he was quick to point out that although man was "the most surprising and admirable" being on the earth, man combined in himself the very worst qualities as well as the very best.[117] The extent of Lamarck's pessimism about his own species was reflected in his development of a theme he had treated briefly some three decades earlier: man's threat to nature's balance. In the 1790s he had commented on how man could destroy a fertile country by the unwise cutting of forests. In the early 1800s he had gone on to identify man as the cause of the extinction of certain large animals, if such extinctions had indeed occurred. In 1817 these views, and no doubt some of the bitterness of the neglected naturalist-philosopher, spilled forth in one of the very earliest prophecies of humanly-wrought global ecological disaster:

> By his egoism too short-sighted for his own good, by his tendency to revel in all that is at his disposal, in short, by his lack of concern for the future and for his fellow man, man seems to work for the annihilation of his means of conservation and for the destruction of his own species. In destroying everywhere the large plants that protect the soil in order to secure things to satisfy his greediness of the moment, man rapidly brings about the sterility of the ground on which he lives, dries up the springs, and chases away the animal that once found their subsistence there. He causes large parts of the globe that were once very fertile and well populated in all respects to become dead, sterile, uninhabitable, and deserted. Neglecting always the words of experience, abandoning himself to his passions, he is perpetually at war with his own kind, destroying them everywhere and under all pretexts, so that one sees formerly great populations become more and more diminished. One could say that he is destined to exterminate himself, after having rendered the globe uninhabitable.[118]

Lamarck's life was beset with difficulties: his parents' desire that he be a priest, the injury that forced him to retire from the army, the death of three successive wives, the loss through a bad investment of the small amount of money he had inherited, the progressive weakening and eventually total failure of his eyesight, and the ridicule or neglect of the ideas he cherished most. The moments that might have been triumphs for him were clouded by his personal problems. A list of the professors of the Muséum d'Histoire Naturelle in 1794 gives a capsule summary of Lamarck's status at the moment he began embarking on his new career: "age 50, married for the second time, wife pregnant, six children, professor of zoology of the insects, worms, and microscopic animals."[119] In 1809, the year in which his *Philosophie zoologique* appeared, he was offered the chair of zoology at the Faculté des Sciences, but he turned it down, explaining: "having consulted the extreme weakness of my physical forces, as well as the habitual state of my bad health, I see myself definitely forced not to accept the honor."[120] Lamarck found solace for his woes in the study of nature. Happiness, he maintained, could be achieved by recognizing one's place in nature and by living in harmony with nature's laws.

It is not surprising that Lamarck regarded Rousseau as "the most profound of our moralists."[121] What he admired in Rousseau was his basic passion for nature and awareness of the dangers of contradicting nature's laws. "Sharing [Rousseau's] sentiments," Lamarck observed, "I dare say that of all our knowledge the most useful to us is that of nature."[122] To study nature was not just to distinguish and classify her productions, but rather to know nature herself: her power, her laws, and "the constant *marche* that she follows in all that she does."[123] By studying nature, Lamarck concluded, man "will conform more easily to nature's laws . . . , he will be able to elude ills of all kinds, and finally, he will derive the greatest advantages."[124]

Voltaire and certain other philosophers were wrong, Lamarck noted, in "regarding as ills and disorders what belongs essentially to the nature of things, what is only the result of a general and constant order of changes, destructions, and renewals . . ."[125] Rousseau, Lamarck observed, had refuted Voltaire by sentiment: "He would have been still more victorious had he recognized this general order that the powerful AUTHOR of all that exists has instituted in the diverse parts of the universe."[126] All natural phenomena, Lamarck maintained, are subject to law. "The word *chance*," he explained, "expresses only our ignorance of causes."[127]

Having once developed the conviction that the state of everything within the domain of nature was to be explained by the "order of things" that constituted nature, Lamarck acknowledgd only too rarely that a problem relating to living things was beyond his grasp. For example, there is no suggestion in his major biological writings of a mechanism by which characters acquired in an individual's lifetime might be passed along to subsequent generations. Indeed there is no indication that he had ever considered the question seriously. Only in the very last exposition of his evolutionary views did he admit that the means by which generation takes place still remained a mystery to him: "Although we have followed attentively the different kinds of reproduction observed, the order of phenomena they present, and the conditions they require in each particular case, it will perhaps be a long time

An unpublished sketch of Lamarck drawn the year before his death. Courtesy of the Library of the Museum of Comparative Zoology, Harvard University.

before we are able to grasp the true mechanism nature employs in the execution of these phenomena."[128] Significantly, he did not hesitate to add: "We are quite convinced, however, that it is not beyond our power to arrive there."[129]

The conviction that all observable phenomena were to be explained wholly by natural causes formed the cornerstone of Lamarck's writings on theoretical biology in the nineteenth century. This was by no means a conviction shared by all of Lamarck's contemporaries. Nor had Lamarck himself always subscribed to it. But it was the kind of leap of faith that Lamarck was prone to make. In 1815 he argued that man's positive knowledge was limited to bodies, to the "properties, faculties, and phenomena that these bodies present," and to "*nature,* which changes, diversifies, destroys, and perpetually renews them."[130] All bodies, he claimed, owed to nature "their existence, their state, their properties, their faculties, and all the changes they undergo; ... [they] are all truly her productions."[131] This philosophical decision was not something, however, that Lamarck had arrived at independently of his zoological studies. Instead, it was intimately related to the way his understanding of the nature of life developed as he contemplated the astonishing variety of forms and faculties displayed by the invertebrates. When he assumed his professorship at the museum he regarded life as an inconceivable, vital principle, but only a few years later he dropped that view completely. By 1797 he was treating animal life as well as plant life simply as a kind of motion. It was at that point that he allowed that "without rejecting anything that pertains to religious belief, nor which may be consoling for the virtuous man to believe in, I will say that [the consideration of the soul] is absolutely foreign to my subject; because the immortal soul of man and the perishable soul of beasts etc. cannot be known to me physically."[132] The natural world, Lamarck seemed prepared to assert, was ultimately comprehensible to man. Nature's method of creating and developing organs seemed no problem to him: it was simply a matter of fluids carving out channels in cellular tissues. And by 1800 nature's method of producing the simplest forms of life also appeared explicable to him: the simplest forms of life arose initially through the action of subtle fluids on small mucilaginous or gelatinous masses when conditions were right. Lamarck no longer felt, as he had in 1794, that the cause of the existence of organic beings and of that which constituted life within them would be forever incomprehensible to man. Since he had already believed for a long time that the phenomena of the inorganic world, including the origin of the different sorts of minerals, were wholly explicable by natural causes, as of 1800 the whole natural world lay before him, waiting to be made intelligible.

Confident in the ultimate explicability of natural phenomena, Lamarck was too ready to believe that the basic explanations of these phenomena were close at hand. He believed he could account for a host of physico-chemical phenomena with the idea that the tendency of all compounds was to decompose. He thought he could explain the major features of the earth's surface by the effects of water moving upon it. He expected to find a regular pattern in the changes in the weather that could be correlated with the presumed action of the sun and the moon on the earth's atmosphere. He sought to understand the natural order of classification, and he determined that the "power of life" and the influence of the environment were the only causes he really needed. There can be no doubt that Lamarck oversimplified problems and neglected the kind of detail that his contemporaries had come to demand in scientific treatises. But it was his broad faith in the comprehensibility of natural phenomena and his general concern with the "large" rather than the "small" facts that enabled him to construct his science of biology and his theory of evolution in the first place.

Many of the elements of the evolutionary theory Lamarck constructed were not novel. Nonetheless, the theory itself was in structure and scope significantly different from anything that had come before it. And though Lamarck did not come up with an entirely satisfactory explanation of the facts of organic diversity; though he did not succeed in accounting for organic phenomena in the way Newton accounted for the motions of planets, the falling of bodies on the earth, and the ebb and flow of the tides; nevertheless, in his general idea of organic evolution, he embraced a notion that would eventually become the great unifying idea of the biological sciences.

Lamarck is not best understood as a man ahead of his time. The role that he saw himself filling as a naturalist-philosopher, and the problems that led him to believe in evolution in the first place, clearly situate him in the last quarter of the eighteenth century and the early years of the nineteenth century. He was a speculator and a builder of scientific systems at a time when a strongly empirical methodology was taking over the natural sciences in France. Simultaneously, he was a professional naturalist who brought his expertise to bear on the particular issues that enlivened natural history in his day. Among these, the reality of extinction and the natural way to classify nature's productions were the issues that played the most immediate roles in the inspiration of his evolutionary views. Not all his thoughts on living things and their origins were felicitous. What is more, as a scientific strategist Lamarck was inept. He was, as Tourlet observed, a bold leaper of intellectual precipices who failed to appreciate the difficulties others might have in following him. Viewing Lamarck's evolutionary theory in the context of the science of the time, one cannot help but feel sympathy toward its author. Lamarck's achievements and failures as an evolutionary theorist were inextricably intertwined.

NOTES

BIBLIOGRAPHY

INDEX

Notes

Most French documents from 1794 to 1807 bear dates corresponding to the Republican calendar, whose first year, *an* 1, began with the founding of the Republic on 22 September 1792. A year was divided into twelve months of thirty days each, with five or six "complementary days" at the end of the year reserved for national holidays. Fall months had the suffix *-aire,* winter months *-ose,* spring *-al,* and summer *-or:* Vendémiaire, Brumaire, Frimaire, Nivôse, Pluviôse, Ventôse, Germinal, Floréal, Prairial, Messidor, Thermidor, Fructidor.

The titles of some of Lamarck's works have been abbreviated in the notes. Full titles can be found in the bibliography.

Botanique	*Encyclopédie méthodique: botanique,* 8 vols. (Paris, 1783–1817).
Dictionnaire	*Nouveau dictionnaire d'histoire naturelle,* 36 vols. (Paris, Deterville, 1816–1819).
"Espèce" (1786)	"Espèce," *Encyclopédie méthodique: botanique, 2* (1786), 395–396.
"Espèce" (1817)	"Espèce," *Nouveau dictionnaire d'histoire naturelle, 10* (1817), 441–451.
Faits physiques	*Recherches sur les causes des principaux faits physiques,* 2 vols. (Paris, 1794).
Flore	*Flore françoise,* 3 vols. (Paris, 1778 [1779]).
HNASV	*Histoire naturelle des animaux sans vertèbres,* 7 vols. (Paris, 1815–1822).
Inédits	*Inédits de Lamarck,* ed. Max Vachon, Georges Rousseau, and Yves Laissus (Paris, 1972).

MPHN	*Mémoires de physique et d'histoire naturelle* (Paris, 1797).
"Nature"	"Nature," *Nouveau dictionnaire d'histoire naturelle, 21* (1818), 363–399.
PZ	*Philosophie zoologique,* 2 vols. (Paris, 1809).
RCV	*Recherches sur l'organisation des corps vivans* (Paris, 1802).
SASV	*Système des animaux sans vertèbres* (Paris, 1801).

Introduction

1. RCV, 208. The example was promoted to the main text in Lamarck's *Philosophie zoologique* (Paris, 1809), *1,* 256–257.

2. On the Kammerer affair see Arthur Koestler, *The case of the midwife toad* (New York, 1971), and the important historical and biological criticisms of Koestler's book offered by Stephen Jay Gould, "Zealous advocates," *Science, 176* (1972), 623–625.

3. *The situation in biological science,* Proceedings of the Lenin Academy of Agricultural Sciences of the USSR (Moscow, 1949), 614.

4. On the Lysenko affair see Zhores A. Medvedev, *The rise and fall of T. D. Lysenko* (New York, 1969); David Joravsky, *The Lysenko affair* (Cambridge, Mass., 1970); and Loren R. Graham, *Science and philosophy in the Soviet Union* (New York, 1972).

5. Lamarck, "Froment," *Botanique, 2* (1788), 558.

6. RCV, 202.

7. Lamarck, "La biologie," ed. Pierre-P. Grassé, *Revue scientifique, 5* (1944), 270.

8. Ibid.

9. See especially Alpheus S. Packard, *Lamarck, the founder of evolution: his life and work* (New York, 1901). The return to a "Lamarckian" understanding of evolution is also signalled in the major biography of Lamarck: Marcel Landrieu, *Lamarck, le fondateur du transformisme* (Paris, 1909).

10. Charles Coulston Gillispie, "Lamarck and Darwin in the history of science," in Bentley Glass, Owsei Temkin, and William L. Strauss, Jr. (eds.), *Forerunners of Darwin: 1745–1859* (Baltimore, 1959), 265–291 (the quote is from p. 286). See also Gillispie, "The formation of Lamarck's evolutionary theory," *Archives internationales d'histoire des sciences, 9* (1956), 323–338, and his *The edge of objectivity: an essay in the history of scientific ideas* (Princeton, 1960). Gillispie's writings are extremely stimulating, and his view of Lamarck as an essentially "pre-scientific" character contains a certain element of truth. Two major aspects of his interpretation of Lamarck's work cannot, however, stand up to historical

scrutiny. The first is that Lamarck's idea of organic mutability was a direct outgrowth of his idea of mineral mutability. The second is that Lamarck endorsed a basically romantic philosophy of nature.

11. John C. Greene, *The death of Adam: evolution and its impact on Western thought* (Ames, Iowa, 1959), provides a good overview of Lamarck's evolutionary theory and a more accurate idea of Lamarck's philosophy of nature than that offered by Gillispie. Jean-Paul Aron, "Les circonstances et le plan de la nature chez Lamarck," *Essais d'épistémologie biologique* (Paris, 1969), 83–98, makes several interesting observations on some of the apparent inconsistencies in Lamarck's evolutionary theorizing. M. J. S. Hodge, "Lamarck's science of living bodies," *British Journal for the History of Science, 5* (1971), 323–352, offers a balanced view of the central concerns of Lamarck's biology and makes the important point that Lamarck was not concerned so much with species change as he was with the successive production of living things from the simplest to the most complex. See also Hodge's "Species in Lamarck," in J. Schiller (ed.), *Colloque international "Lamarck"* (Paris, 1971), 31–46. Leslie J. Burlingame, "Lamarck," *Dictionary of scientific biography, 7* (1973), 584–594, provides the best brief overview of the whole of Lamarck's scientific thought. J. Schiller, "Physiologie et classification dans l'oeuvre de Lamarck," *Histoire et biologie, 2* (1969), 35–57, and "L'échelle des êtres et la série chez Lamarck," in Schiller (ed.), *Colloque international "Lamarck,"* 87–103, has paid special attention to the connections between Lamarck's evolutionary thought and his ideas about classification and physiology. Gabriel Gohau, "Le cadre minéral de l'évolution Lamarckienne," in Schiller (ed.), *Colloque international "Lamarck,"* 105–133, has examined the role of Lamarck's mineralogy as a model for his later theory of organic change. Ernst Mayr, "Lamarck revisited," *Journal of the History of Biology, 3* (1970), 275–298 offers an evolutionary biologist's perceptive reading of Lamarck's *Philosophie zoologique.* Other recent writings on Lamarck include I. M. Poliakov, *J.-B. Lamarck and the theory of the evolution of the organic world,* in Russian (Moscow, 1962); Franck Bourdier, "L'homme selon Lamarck," in Schiller (ed.), *Colloque international "Lamarck,"* 137–159, and "Lamarck et Geoffroy Saint-Hilaire face au problème de l'évolution biologique," *Revue d'histoire des sciences, 25* (1972), 311–325; Frans Stafleu, "Lamarck: the birth of biology," *Taxon, 20* (1971), 397–442; and J. Brémond and J. Lassertisseur, "Lamarck et l'entomologie," *Revue d'histoire des sciences, 26* (1973), 231–250. Despite these recent analyses, perhaps the richest treatment of Lamarck's thought is still to be found in Henri Daudin's classic studies, *De Linné à Jussieu: méthodes de classification et idée de série en botanique et en zoologie (1740–1790)* (Paris, n.d. [1926]), and *Cuvier et Lamarck: les classes zoologiques et l'idée de série animale (1790–1830),* 2 vols. (Paris, 1926).

At the same time that new attention has been given to Lamarck's work, new Lamarck materials have been brought to light. For an assessment of current knowledge concerning Lamarck manuscripts, see M. Vachon, "A propos des manuscrits de Lamarck conservés à la bibliothèque centrale du Muséum national d'Histoire naturelle de Paris (note préliminaire)," *Bulletin du Muséum National d'Histoire Naturelle, 39* (1967), 1023–1027, and Max Vachon, Georges Rousseau, and Yves Laissus, "Liste complète des manuscrits de Lamarck conservés à la bibliothèque centrale du Muséum national d'histoire naturelle de Paris," *Bulletin du Muséum National d'Histoire Naturelle, 40* (1968), 1093–1102. An important portion of the Lamarck manuscripts at the Muséum has recently been published in Max Vachon, Georges Rousseau, and Yves Laissus (eds.), *Inédits de Lamarck: d'après les manuscrits conservés à la bibliothèque centrale du Muséum National d'Histoire Naturelle de Paris* (Paris, 1972). For a review of this volume see Richard W. Burkhardt, Jr., *British Journal for the History of Science, 7* (1974), 192–194. Some little known or unknown publications by Lamarck have also been discovered. M. J. S. Hodge has found a concise summary of Lamarck's metaphysical views that was believed to be lost: "La métaphysique de Lamarck d'après un opuscule retrouvé," *Revue d'histoire des sciences, 26* (1973), 223–229. I have discovered four meteorological articles by Lamarck that have not been cited by Lamarck's earlier biographers: "Considérations sur les observations météorologiques comparatives, et sur leur application," *Annales de statistique, 3* (1802), 58–72; "Lettre du cit. J.-B. Lamarck, à L.-J. P. Ballois," *Annales de statistique, 3* (1802), 78–82; "Sur les relations qui existent entre le météorologie-statistique et la météorologie générale, et sur les moyens d'avancer les progrès de l'une et de l'autre," *Annales de statistique, 3* (1802), 300–317; "Sur les observations et le but de toute météorologie-statistique," *Annales de statistique, 4* (1802), 129–134. These various additions to the corpus of Lamarck's writings provide some added insights into Lamarck's scientific thought. By and large, however, they do not force a revision of the image of Lamarck that can be gained from his better-known publications. A final source of some value concerning the influences upon Lamarck is the catalogue of the books in his library that were put up for auction following his death: *Catalogue des livres de la bibliothèque du feu M. le Chevalier J.-B. de Lamarck* (Paris, 1830).

12. Lamarck, "De l'influence de la lune sur l'atmosphère terrestre," *Journal de physique, 46* (1798), 429. On Lamarck's thoughts on the incompatibility between physical and mental activity, see for example RCV, 128, and PZ, *1*, 248; *2*, 409–410.

13. Landrieu, *Lamarck* (Paris, 1909), is the standard bibliographical source. Also employed here are Burlingame, "Lamarck," and Franck Bourdier (and Michael Orliac, collaborator), "Esquisse d'une chronologie de la vie de Lamarck," unpublished ms, 3rd section, École Pratique des Hautes Études, 22 June 1971.

14. Auguste de Lamarck's comments are published by Landrieu, *Lamarck,* 79.

15. James Edward Smith, *A sketch of a tour on the continent, in the years 1786 and 1787,* 3 vols. (London, 1793), *1,* 123–124, speaks of Lamarck's rudeness to him.

16. See for example Cuvier, *Tableau élémentaire d'histoire naturelle* (Paris, an 6 [1798]), vii; Desfontaines, *Tableau de l'école de botanique du jardin du roi,* 2nd ed. (Paris, 1815), v; Faujas de Saint-Fond, *Histoire de la montagne de Saint-Pierre de Maestricht* (Paris, [1799]), 126–136; Lamouroux, *Histoire des polypes coralligènes flexibles* (Caen, 1861), viii.

17. Augustin Pyramus de Candolle, *Mémoires et souvenirs de Augustin Pyramus de Candolle, écrites par lui même et publié par son fils* (Genève, 1862), 186–187, describes how Lamarck was pressured into not supporting him.

18. Cited in Landrieu, *Lamarck,* 22.

19. Isidore, Bourdon, *Illustres médecins et naturalistes des temps modernes* (Paris, 1844), 343–344.

20. The memoir was Lamarck's "Mémoire sur les variations de l'état du ciel," later published in the *Journal de physique, 56* (1802), 114–138. The incident is referred to in a letter from Étienne Geoffroy Saint-Hilaire to Cuvier, Institut de France, Fonds Cuvier, MS 3225 (12). Geoffroy felt that Laplace's attack on Lamarck was excessive.

21. François Arago, *Histoire de ma jeunesse,* in *Oeuvres de François Arago,* 17 vols. (Paris and Leipzig: 1854–1862), *1,* 94. Lamarck complained of Napoléon's disenchantment with his meteorological studies in "Météorologie," *Dictionnaire, 20* (1818), 475–476.

22. Charles Darwin felt that his grandfather's *Zoonomia,* 2 vols. (London, 1794, 1796) anticipated Lamarck's evolutionary theory, and he once hinted in a letter to Huxley that the similarities between his grandfather's views and those of Lamarck might not have been wholly fortuitous (see Francis Darwin [ed.], *More letters of Charles Darwin,* 2 vols. [London, 1903], *1,* 125). But there is no evidence to support the assumption that Erasmus Darwin was the source of Lamarck's evolutionary ideas. The *Zoonomia* was not translated into French until 1810, and Lamarck owned neither the French nor the original English version of it. A very brief notice of the first volume of the first English edition appeared in the *Magasin encyclopédique, 1* (1795), no. 4, 558, but this notice made no mention of the thoughts on organic mutability contained in the work, which is scarcely surprising since these thoughts make up only part of one of the forty sections of the book. To the best of my knowledge, in Lamarck's lifetime Erasmus Darwin's name was never mentioned in a French discussion of transformist ideas. It seems likely that Cuvier would have mentioned Erasmus Darwin as a precursor of Lamarck had he known of Darwin's views, and certainly Cuvier would have known of these views

had they been a subject of discussion among French naturalists. As it was, Cuvier found an obscure German named Rodig with whom he frequently compared Lamarck (see Chapter Seven).

Chapter One
Jean-Baptiste Lamarck, Naturalist-Philosopher

1. Lamarck, *Hydrogéologie* (Paris, 1802), 7–8.

2. The memoir was read at the Institute on 15 April 1798 (26 Germinal an 6) and published as "De l'influence de la lune sur l'atmosphère terrestre," *Journal de physique, 46* (1798), 428–435.

3. Lamarck began reading the memoir on 9 February 1799 (21 Pluviôse an 6) and completed it on 19 February 1799 (1 Ventôse an 7). The memoir is identified as being on "les fossiles et l'influence du mouvement des eaux, considérées comme preuves du déplacement continuel du bassin des mers, et de son transport sur les différens points de la surface du globe" in Institut de France, Académie des Sciences, *Procés-verbaux des séances de l'Académie tenues depuis la fondation de l'Institut jusqu'au mois d'août 1835,* 524, 528. Lamarck's observations were reviewed briefly by Lassus, "Sciences physiques," *Magasin encyclopédique, 5* (1799), no. 1, 233–234.

4. Lamarck, "Discours d'ouverture prononcé le 21 floréal an 8," SASV, 1–48.

5. Daniel Mornet, *Les sciences de la nature en France au XVIIIe siècle* (Paris, 1911), 248–249.

6. Yves Laissus, "Les cabinets d'histoire naturelle," in René Taton (ed.), *Enseignement et diffusion des sciences en France au XVIIIe siècle* (Paris, 1964), 659–712 (see p. 666).

7. See Roger Hahn, *The anatomy of a scientific institution: the Paris Academy of Sciences, 1666–1803* (Berkeley, 1971), 87–92; Laissus, "Les cabinets d'histoire naturelle."

8. Laissus, "Les cabinets d'histoire naturelle," 664.

9. Denys de Montfort, *Conchyliologie systématique, et classification méthodique des coquilles . . .* (Paris, 1810), *2,* 96.

10. Laissus, "Les cabinets d'histoire naturelle," 665.

11. See Lamarck, "Herborisations," *Botanique, 3* (1789), 116–117.

12. J. J. Rousseau, *Lettres élémentaires sur la botanique,* in *Oeuvres complètes de J. J. Rousseau, 5* (Paris, 1874), 30.

13. *Revue générale des écrits de Linné,* ed. Richard Pulteney, trans. L. A. Millin de Grandmaison (London, 1789), *2,* 159–160.

14. "Objets d'histoire naturelle recueillis en Hollande," *Décade philosophique, 5* (an 3 [1795]), 535–536. See also Ferdinand Boyer, "Le Muséum d'Histoire Naturelle à Paris et l'Europe des sciences sous la Convention," *Revue d'histoire des sciences, 26* (1973), 251–257.

15. Daubenton, "Histoire naturelle," *Encyclopédie, 8* (1755), 228.

16. Cited in Robert Olby, *Origins of Mendelism* (New York, 1966), 19.

17. Rousseau, *Lettres élémentaires sur la botanique,* 82. Writing to M. de la Tourette in 1769, Rousseau observed: "I am not surprised that you returned from Italy more satisfied with nature than with men; that is what generally happens to good observers, even in climates where it is less beautiful."

18. Mornet, *Les sciences de la nature,* 168.

19. Tessier, "Mémoire sur l'importation et les progrès des arbres à épicerie dans les colonies françoises," *Mémoires de l'Académie des Sciences* (1789), 585–596.

20. On Daubenton see Cuvier, "Éloge historique de Daubenton," *Recueil des éloges historiques,* new ed. 3 vols. (Paris, 1861), *1,* 3–34.

21. Daubenton, "Introduction à l'histoire naturelle," *Encyclopédie méthodique; histoire naturelle, 1* (1782), i.

22. Ibid.

23. Daubenton, "Programme: sur la définition et les limites de l'histoire naturelle," in *Séances des écoles normales, recueillies par des sténographes, et revues par les professeurs,* new ed., 8 vols. (Paris, 1800), *1,* 103–104.

24. Ibid., *1,* 96.

25. "Introduction," *Encyclopédie méthodique; histoire naturelle, 1,* v.

26. Ibid., *1,* iii.

27. "Histoire naturelle," *Encyclopédie, 8* (1755), 226.

28. "Les trois regnes de la nature," *Encyclopédie méthodique: histoire naturelle, 1,* xi.

29. "Sur la nomenclature méthodique de l'histoire naturelle," *Séances des écoles normales, 1,* 427.

30. "Sur la rédaction de l'histoire naturelle," *Séances des écoles normales, 1,* 291.

31. The word "biology" was also used at this time by Burdach (1800) and Treviranus (1802). See Schiller, "Physiologie et classification dans l'oeuvre de Lamarck," 38.

32. Lamarck, *Hydrogéologie,* 8.

33. Lamarck, "La biologie," *Revue scientifique, 5* (1944), 267–276.

34. Unless indicated otherwise, the biographical data given here are taken from Landrieu or Bourdier (see note 13 of the introduction).

35. Correspondence relating to the publication of the *Flore* may be found in *Oeuvres complètes de Buffon,* ed. J.-L. Lanessan, correspondence collected and annotated by J. Nadault de Buffon (Paris, 1884–1885), *14,* 356–360. The source of information on the vote at the Académie des Sciences is given below in note 38.

36. *Flore, 1,* li-lviii.

37. Daudin, *De Linné à Jussieu,* 191.

38. Manuscript of Académie des Sciences, "Collection de ses reglemens et délibérations par ordre de matière," annotated by J. J. Lalande, Florence, Biblioteca Medicea-Laurenziana, Ashburnham-Libri no. 1700, p. 119. According to the same manuscript (p. 32), there was another vote between Lamarck and Descemet at the academy, this time in 1783 for the next higher position, that of associate member. The vote was a tie, and Lamarck was again awarded the position, thanks once more to the support of the Société Royale de Médecine and the political pull of d'Angiviller. I am indebted to Roger Hahn for providing me with this information and the interpretation of why members of the Société Royale de Médecine at the academy supported Lamarck.

39. Frans Stafleu, "L'Héritier de Brutelle, the man and his work," in Charles-Louis L'Héritier de Brutelle, *Sertum Anglicum, 1788,* facsimile ed. (Pittsburgh, 1963), xiii–lxvi (see xix).

40. James Edward Smith, *A sketch of a tour on the continent, in the years 1786 and 1787,* 3 vols. (London, 1793), *1,* 123.

41. Lamarck, *Flore, 2,* iv.

42. This memoir was abstracted in Louis Cotte, *Mémoires sur la météorologie,* 2 vols. (Paris, 1788), *1,* 205–215. Burlingame, "Lamarck," 587, has identified Muséum d'Histoire Naturelle, MS 755-1 as the manuscript of this unpublished memoir.

43. *Faits physiques, 2,* 366.

44. The only exception to this were two brief articles on conchology in the *Journal d'histoire naturelle:* "Observations sur les coquilles, et sur quelques-uns des genres qu'on a établis dans l'ordre des vers testacés," *2* (1792), 269–280; "Sur quatre espèces d'Hélices," *2* (1792), 347–353.

45. Consult *Mémoires de l'Académie Royale des Sciences,* 1779–1793 [1797].

46. Lamarck, "Mémoire sur un nouveau genre de plante nommé Brucea et sur le faux Bresillet d'Amérique," *Mémoires de l'Académie Royale des Sciences* (1784), 342–347; "Mémoire sur les classes les plus convenables à etablir parmi les vegetaux," *Mémoires de l'Académie Royale des Sciences* (1785), 437–453; "Mémoire sur le genre du Muscadier, Myristica," *Mémoires de l'Académie Royale des Sciences* (1788), 148–168.

47. *Faits physiques, 1,* vii–xvi.

48. E.-T. Hamy, "Les derniers jours du Jardin du Roi et la fondation du Muséum d'Histoire Naturelle," *Centenaire de la fondation du Muséum d'Histoire Naturelle. 10 juin 1793-10 juin 1893* (Paris, 1893), 1–162 (see 3–4).

49. Lamarck, *Considérations en faveur du Chevalier de la Marck . . .* (Paris, 1789), and *Mémoire sur le projet du Comité des finances relatif à la suppression de la place de botaniste, attaché au cabinet d'histoire naturelle* (Paris, n.d.). These pamphlets have been reproduced in full by Landrieu, *Lamarck,* 34–39.

50. Landrieu, *Lamarck,* 36.
51. Ibid., 52n.
52. Ibid., 42–51.
53. Ibid., 46.
54. Hamy, "Les derniers jours du Jardin du Roi," 107–129.
55. Ibid., 66–67.
56. Hahn, *Anatomy of a scientific institution,* 176–181.
57. The main sources on the Société d'Histoire Naturelle de Paris are *Actes de la Société d'Histoire Naturelle de Paris, 1* (1792); *Mémoires de la Société d'Histoire Naturelle de Paris, 1* (1799); and Muséum National d'Histoire Naturelle, MS 298–299 (diverse papers relating to the Société d'Histoire Naturelle de Paris), and MS 464, "Procès-verbaux de la Société d'Histoire Naturelle, 27 août 1790 - 18 prairial an 5ᵉ." I am grateful to Roger Hahn for making available to me his notes on the "Procès-verbaux."
58. "Procès-verbaux de la Société d'Histoire Naturelle," 82–83, 103–104, 168–182.
59. Muséum d'Histoire Naturelle, MS 298, "Rapport fait à la Société d'Histoire Naturelle de Paris, sur les moyens d'activer et d'utiliser ses sceances" (signed Lelievre).
60. Hahn, *Anatomy of a scientific institution,* 112–114.
61. Membership lists of the respective societies are given in *Actes de la Société d'Histoire Naturelle de Paris, 1* (1792), and Paris, Bibliothèque Mazarine, MS 4441, "Registre des Procès-verbaux des Séances de la Société Linnéenne de Paris."
62. Hahn, *Anatomy of a scientific institution,* 179–180.
63. Lamarck, "Instructions aux voyageurs autour du monde sur les observations les plus essentielles à faire en botanique, *Bulletin de la Société Philomathique, 1* (1791), 8.
64. Lamarck, "Sur les ouvrages généraux en histoire naturelle; et particulièrement sur l'Édition du *Systema Naturae* de Linneus, que M. J. F. Gmelin vient de publier," *Actes de la Société d'Histoire Naturelle de Paris, 1* (1792), 81–85.
65. "Procès-verbaux de la Société d'Histoire Naturelle," 104–107.
66. Ibid., 123.
67. Ibid., 158.
68. Muséum National d'Histoire Naturelle, MS 298, "Extrait des Registres de la Société d'Histoire Naturelle de Paris," art. 3.
69. Henry Guerlac, "Some aspects of science during the French Revolution," *Scientific Monthly, 80* (1955), 95, and Hahn, *Anatomy of a scientific institution,* 169, list Lamarck as a member of this society on the basis of Augustin Challamel, *Les clubs contre-révolutionnaires* (Paris, 1885), 409. The Lamarck listed by Challamel, however, proves to have been Auguste-Marie-Raymond, prince d'Aremberg, comte de Lamarck (1753–1833), a friend of Mirabeau.

70. Stafleu, "L'Héritier," xxxi.

71. Landrieu, *Lamarck,* 58–63.

72. *Faits physiques, 1,* v–vi.

73. The countermanding order on Lamarck's behalf was made April 17, 1794. See F.-A. Aulard, *Recueil des actes du Comité de Salut Public* (Paris, 1899), *12,* 640.

74. Landrieu, *Lamarck,* 71.

75. Ibid., 72–73.

76. On the Société Philomathique see Hahn, *Anatomy of a scientific institution,* 178–179, 266–268.

77. "Procès-verbaux de la Société d'Histoire Naturelle," 216–220.

78. Mornet, *Les sciences de la nature,* 73–107.

79. Condillac, cited in Mornet, *Les sciences de la nature,* 104–105.

80. Cited in Mornet, *Les sciences de la nature,* 110.

81. Lamarck, *Flore, 1,* cxviii. See also "Genres," *Botanique, 2* (1788), 633.

82. Lamarck's letter, dated Paris, le 4 Vendémiaire an 3 (25 September 1794), is cited in full by Landrieu, *Lamarck,* 95–97.

83. Lamarck, *Annuaire météorologique pour l'an XIV* (Paris, n.d. [1805]), 5, 99.

84. RCV, vi–vii.

85. PZ, 1, 370.

86. Cited in Mornet, *Les sciences de la nature,* 98.

87. Muséum National d'Histoire Naturelle, MS 756, "Physique terrestre," 1st cahier, p. 3. The passage is from the introductory discourse, which is entitled: "Discours contenant une discussion critique sur les théories physiques en général, sur celles maintenant établies, sur les moyens pris pour les maintenir, enfin sur les difficultés d'opérer des rectifications dans les écarts où l'on s'est jetté." The original text of the passage cited is as follows: "C'est à présent un mérite fort estimé que de ne s'occuper qu'à recueillir des faits, on doit en rechercher de toutes parts; on doit les considérer tous isolément; enfin on doit se circonscrire partout dans les plus petits détails; cette marche seule, dit-on est estimable.

"Pour moi je pense qu'il peut etre maintenant utile de rassembler les faits recueillis, et de s'efforcer à les considerer dans leur ensemble, afin d'en obtenir les résultats généraux les plus probables. Celui qui conclueroit que dans l'étude de la nature, nous devons toujours nous borner à amasser des faits; ressembleroit à un architecte qui conseilleroit toujours de tailler des pierres, de préparer des mortiers, des bois, des ferrures, &c et qui n'oseroit jamais employer ces matériaux pour construire un édifice."

88. *Hydrogéologie,* 5–6.

89. Ibid., 7.

90. Lamarck, "Considérations sur quelques faits applicables à la théorie du globe, observés par M. Péron dans son voyage aux terres aus-

trales, et sur quelques questions géologiques qui naissent de la connoissance de ces faits," *Annales du Muséum d'Histoire Naturelle, 6* (1805), 38–39.

91. Lamarck, *Réfutation de la théorie pneumatique* (Paris, 1796), 3. For similar comments, see *Hydrogéologie,* 103, 122, 159n, and 164. For comments regarding the neglect of his meteorological work, see "Sur les variations de l'état du ciel..." *Journal de physique, 56* (1802), 138; his *Annuaire météorologique pour l'an 1808* (Paris, n.d. [1807]), and "Météorologie," *Dictionnaire, 20* (1818), 474–477.

92. Lamarck, *Réfutation,* 78–481 contrasts his pyrotic theory page by page with the second edition of Fourcroy's *Philosophie chimique* (Paris, an 3 [1795]).

93. "Nouvelles littéraires," *Magasin encyclopédique, 2* (1796), no. 4, 406–407. At the public session of the Société de Santé de Nancy, a citizen Nicolas read a critique of Lamarck's *Réfutation de la théorie pneumatique.* Nicolas cited experimental evidence in support of the chemical theory of Lavoisier, Guyton-Morveau, Fourcroy, Berthollet, and Chaptal.

94. MPHN, 410.

95. For the title that Lamarck initially intended to use for this work, see Chapter Four, note 3.

96. "Discours prononcé à la Société Philomathique, le 23 floréal an 5," 3. A copy of this discourse is bound together with the copy of Lamarck's MPHN at the University of Illinois, Urbana-Champaign.

97. Lamarck resigned from the Société Philomathique on 20 August 1797 (3 fructidor an 5).

98. Landrieu, *Lamarck,* 20–21.

99. Smith, *A sketch of a tour on the continent, 1,* 123–124.

100. When Lamarck presented his physical *Recherches* to the academy in 1780 he took care to have the perpetual secretary affix his paraph to the manuscript in order to assure for Lamarck a record of his priority. On another occasion he communicated a meteorological observation to the first class of the institute and had the observation inserted in the *procès verbal* "pour prendre date à ce sujet" (*Procès-verbaux des séances de l'Académie, 1,* 63). More significant were the priority concerns regarding the reform of invertebrate classification, which caused a certain amount of friction between Lamarck and Cuvier (see Chapter Five).

101. Muséum d'Histoire Naturelle, MS 756, "Physique terrestre," 1st cahier, p. 11. The original is as follows: "celui dont on parle c'est un homme qui ne scait rien, qui ne connois pas les faits, qui n'a jamais fait d'exp.[ce]"

102. Ibid. The original is as follows: "l'hist. nous fournit quantité de traits qui nous apprenent tout ce qu'ont eû a souffrir des h. du l.[er] merite pour avoir osé de bonne foi cherché la verité, au lieu de se plier sous l'autorité dominant, au lieu de se soumettre à l'opinion en crédit.

"mettons ici en parallele le savant ord.[re] celui qui de son vivant jouit de tout l'avantage d'une gr. reputation, et l'homme rare toujours inconnu, qui cherche sincerement la verité et la prefere à tous les autres g. de jouissances—vous verrez dans le 1er sous un ext. de modestie qu'il n'a pas un homme suffisant, plein d'orgueil, hardie, toujours decidant d'une maniere breve et tranchie toutes les questions, ayant touj. à la b. cela est demontré. cet h. n'est presque jamais chez lui; on le rencontra dans toutes les belles assemblées, il est membre de tous les corps considerés, chargé de tous les emplois, de toutes les fonctions, affecte" [here the manuscript ends].

103. Augustin Pyramus de Candolle, *Mémoires et souvenirs,* 44.

104. RCV, 126–127.

Chapter Two
The Background to Lamarck's Biological Thought

1. Michel Foucault, *The order of things: an archeology of the human sciences* (New York, 1971), 127–128.

2. Cuvier, *Leçons d'anatomie comparée,* 1 (Paris, an 8 [1800]), iv.

3. Lamarck, *Flore, 1,* 5.

4. Lamarck, "Discours d'ouverture, prononcé le 21 floréal an 8," SASV, 1–48.

5. Ibid., 1–2.

6. Lamarck, "Botanique," *Botanique, 1* (1783), 439–449.

7. Ibid., 442.

8. Ibid., 443.

9. Lamarck, *Inédits,* 160.

10. Daudin, *De Linné à Jussieu,* chapters three and four.

11. James L. Larson, *Reason and experience: the representation of natural order in the work of Carl von Linné* (Berkeley, 1971).

12. Buffon, "De la manière d'étudier et de traiter l'histoire naturelle," in *Oeuvres philosophiques de Buffon,* ed. Jean Piveteau (Paris, 1954), 11.

13. Antoine de Jussieu, "Exposition d'un nouvel ordre de plantes adopté dans les démonstrations du Jardin Royal," *Mémoires de l'Académie Royale des Sciences* (1774), 175–197.

14. Ibid., 197.

15. Lamarck, *Flore, 1,* 1.

16. Ibid., *1,* 2.

17. Lamarck, "Espèce" (1786), 395.

18. Lamarck, *Faits physiques, 2,* 378.

19. Ibid., *2,* 378–379.

20. On this distinction and the distinction between plants and animals, see Daudin, *De Linne à Jussieu,* 177–183. On Daubenton, see

Encyclopédie méthodique: histoire naturelle des animaux, 1, iii; *Séances des écoles normales, 1,* 426, *4,* 3–13.

21. Desfontaines, "Histoire naturelle: cours de botanique élémentaire et de physique végétale; discours d'ouverture," *La Décade philosophique, 5* (an 3 [1795]), 450. Lamarck, "Discours préliminaire pour le cours de l'an six," *Inédits,* 162.

22. Cited by Daudin, *De Linné à Jussieu,* 180, fn. 4.

23. Daudin, *De Linné à Jussieu,* 182–183.

24. Daubenton, "Sur la nomenclature méthodique de l'histoire naturelle," *Séances des écoles normales, 1,* 418.

25. Lamarck, "Classes," *Botanique, 2* (1786), 33.

26. Lamarck, *Faits physiques, 2,* 306.

27. Lamarck, MPHN, 289–290.

28. For his ideas on the chain of being Bonnet continues to be cited as a forerunner of Lamarck, as, for example, in P. E. Pilet, "Charles Bonnet," *Dictionary of scientific biography, 2* (1970), 286. Daudin, *De Linné à Jussieu* and J. Schiller, "L'échelle des êtres et la série chez Lamarck", have both insisted that Lamarck's thinking and that of Bonnet were markedly dissimilar.

29. Cited by Daudin, *De Linné à Jussieu,* 113, fn. 2.

30. Cited by Daudin, *De Linné à Jussieu,* 112.

31. Jussieu, "Exposition d'un nouvel ordre de plantes," 192.

32. Lamarck, *Flore, 2,* iv.

33. Daudin, *De Linné à Jussieu* and *Cuvier et Lamarck.*

34. Daudin, *Cuvier et Lamarck, 2,* 111, 155.

35. Lamarck, SASV, 12.

36. Lamarck only once in his writings mentioned Leibniz's name, and then only to refer to Leibniz's "rare and superior mind." *Faits physiques, 1,* 10.

37. Lamarck, MPHN, 318.

38. *Oeuvres philosophiques de Buffon,* 7–26.

39. Lamarck, "Genres," *Botanique, 2* (1788), 632.

40. Lamarck, *Flore, 1,* xci (misnumbered as cxi).

41. Ibid., *1,* xc-xci.

42. Ibid., *1,* xc.

43. Ibid., *1,* xci.

44. Ibid., *1,* xciii.

45. Ibid., *1,* xciv.

46. Ibid., *1,* cxiii.

47. Lamarck, "Classes," *Botanique, 2* (1786), 29–36. Except for the additional discussion of minerals in this article, the article is virtually the same as Lamarck's "Mémoire sur les classes . . .", *Mémoires de l'Académie Royale des Sciences* (1785), 437–453.

48. Lamarck, "Classes," *Botanique,* 30.

49. Ibid., 29.

50. Ibid., 33. A. L. de Jussieu had pointed out the analogy between the plant and the animal kingdoms in "Examen de la famille des Renoncules," *Mémoires de l'Académie Royale des Sciences* (1773), 214–240.

51. Daudin, *De Linné à Jussieu,* 105–110.

52. Lamarck, *Flore, 1,* lxxxviii-lxxxix.

53. Lamarck, "De ce qui reste à faire pour donner à la botanique le degré de perfection dont elle ne peut se passer," *Inédits,* 49.

54. Lamarck, HNASV, *1,* 461.

55. Johann Hermann endorsed the idea of reticular arrangement. He also mentioned the geographical map analogy in his *Tabulae affinitatum animalium* (Strasbourg, 1783) and Lamarck cites him in this regard (RCV, 40). Pallas also advanced the idea of reticular arrangement. On Hermann and Pallas see Daudin, *De Linné à Jussieu,* 163–166, 170–173.

56. Daubenton, "Sur les rapports que l'on a recherchés entre les corps bruts et les corps organisés," *Séances des écoles normales, 4,* 3–13, and "Sur la nomenclature méthodique de l'histoire naturelle," *Séances des écoles normales, 1,* 425–444, esp. 431 and 433, where the arrangement suggested in the *Flore françoise* is criticized.

57. Lamarck, SASV, 17n.

58. Lamarck, *Flore, 1,* 2; "Botanique," *Botanique, 1,* 441; *Faits physiques, 2,* 184–314.

59. Lamarck, *Faits physiques, 2,* 26n.

60. Ibid., *2,* 286.

61. Ibid., *2,* 185.

62. Ibid., *2,* 186–187.

63. Ibid., *2,* 190.

64. Ibid., *2,* 188–189.

65. Ibid., *2,* 188–219. Lamarck held this for plants as well as for animals. See *Flore, 1,* 103.

66. On Boerhaave and Haller consult Thomas S. Hall, *Ideas of life and matter,* 2 vols. (Chicago, 1969), *1,* 367–408. On Hoffmann, see Lester King, *The growth of medical thought* (Chicago, 1963) and "Stahl and Hoffmann: a study in eighteenth century animism," *Journal of the History of Medicine, 19* (1964), 118–130.

67. D'Aumont, "Génération," *Encyclopédie, 7* (1757), 559–574 (esp. 559). Bonnet, *Considérations sur les corps organisés,* in Charles Bonnet, *Oeuvres d'histoire naturelle et de philosophie, 5* (Neuchatel, 1779), 167–168.

68. Buffon, "De la vieillesse et de la mort," *Histoire naturelle de l'homme,* in *Oeuvres complètes de Buffon,* ed. Lamouroux and A. G. Desmarest, *13* (Paris, 1828), 174–210.

69. Cited by Lester King, *The growth of medical thought,* 163.

70. Baron Albertus Haller, *First lines of physiology,* reprint of 1786 edition, introduction by Lester S. King (New York and London, 1966), 244.

71. Ibid., 246.

72. Lamarck, *Faits physiques, 2,* 220–223.

73. Ibid., *2,* 244.

74. Ibid., *2,* 190.

75. Lamarck, *Flore, 1,* 2.

76. Ibid., *1,* 3.

77. Ibid., *1,* 209–211.

78. Ibid., *1,* 220; *Faits physiques, 2,* 211–213.

79. Lamarck, "Acacia," *Botanique, 1* (1783), 8–22.

80. Ibid., 17–18.

81. Lamarck, "Classes," *Botanique, 2* (1786), 30.

82. Lamarck, "Irritabilité," *Botanique, 3* (1789), 308.

83. Albrecht von Haller, "A dissertation on the sensible and irritable parts of animals," intro. by Owsei Temkin, *Bulletin of the Institute of the History of Medicine, 4* (1936), 651–699 (see 658–659).

84. Ibid., 692.

85. See Philip C. Ritterbush, *Overtures to biology: the speculations of eighteenth-century naturalists* (New Haven, 1964), chapter two, and Robert E. Schofield, *Mechanism and materialism: British natural philosophy in an age of reason* (Princeton, 1970), chapter eight.

86. Cited in Lester King, *The growth of medical thought,* 168.

87. L'Abbé Nollet, *Recherches sur les causes particulières des phénomènes électriques, et sur les effets nuisibles ou avantageux qu'on peut en attendre* (Paris, 1749), 355.

88. Ibid., 355–356.

89. Ibid., 358–424. Joseph Priestley, *The History and present state of electricity, with original experiments,* 3rd ed. (London, 1755), *1,* 472–489, devotes a chapter to "medical electricity." The therapeutic uses of electricity by eighteenth-century physicians are also described by Sigaud de la Fond, *Précis historique et expérimental des phénomènes électriques,* 2nd. ed. (Paris, 1785), 468–504.

90. Toaldo, "Essai de météorologie appliquée à l'agriculture," *Journal de physique, 10* (1777), 249–279, 333–367.

91. Ibid., 254–255.

92. Ibid., 256.

93. Quatremère Disjonval's words to "one of his friends" are reported by Olivier in *Encyclopédie méthodique: histoire naturelle des insectes, 7* (1792), 369. Olivier quotes the ex-academician's discussion of the summer of 1789, which had been very rainy throughout June and July: "le trois ou quatre août, il s'est fait, à deux heures après-midi, une des plus grandes révolutions dans l'atmosphère, qui ait eu lieu peut-être de toute

l'année. Mes Araignées prirent de toute part le mord aux dents, & elles allèrent porter les maîtres brins de nouvelles toiles à des distances énormes par rapport à celles qui précédoient. Au même instant les Coqs & les Pigeons que j'ai sous les yeux, entrerent dans des luttes acharnées. Les vaisseaux spermatiques (de la plus grande irritabilité il est vrai chez moi) me causèrent une érection longue, soutenue, & qui n'étoit provoquée par aucune circonstance de la vue ou de l'imagination. Je ne doutai point que ce ne fut la naissance de l'été." See also Quatremère Disjonval, *De l'Aranéologie, ou sur la découverte du rapport constant entre l'apparition ou la disparition, le travail ou le moins d'étendue des toiles et des fils d'attaches des araignées des differentes espèces* (Paris, an 5 [1797]).

94. Lamarck, "De l'influence de la lune sur l'atmosphère terrestre," *Journal de physique, 46* (1798), 434.

95. Lavoisier, *Traité élémentaire de chimie,* 2 vols. (Paris, 1789), *1,* 7.

96. Franklin, le Roy, De Bory, Lavoisier, and Bailly, "Exposé des expériences qui ont été faites pour l'examen du magnétisme animal," *Histoire de l'Académie Royale des Sciences* (1784), 6–15. On the subject of mesmerism see Robert Darnton, *Mesmerism and the end of the Enlightenment in France* (Cambridge, Mass., 1968).

97. Lavoisier, *Traité élémentaire de chimie, 1,* 4.

98. See Ritterbush, *Overtures to biology,* 51–56; Ingen-Housz, "Lettre . . . au sujet de l'influence de l'électricité atmosphérique sur les végétaux," *Journal de physique, 32* (1788), 321–337; Bertholon, "Nouvelles expériences sur les effets de l'électricité artificielle & naturelle, appliquée aux végétaux," *Journal de physique, 35* (1789), 401–423.

99. Lamarck, *Faits physiques, 2,* 268n.

100. Lamarck, "Classes," Botanique, *2,* 34.

101. Ibid., 35.

102. Ibid., 34.

103. Baumé, *Chymie expérimentale et raisonnée,* 3 vols. (Paris, 1773), *1,* xix.

104. Ibid., 3, 301–334.

105. Lamarck, *Faits physiques, 2,* 306–309.

106. Lamarck, *Flore, 1,* 104, 178.

107. Lamarck, "Forêt," *Botanique, 2* (1788), 519.

108. See Frank N. Egerton, "Changing concepts of the balance of nature," *The Quarterly Review of Biology, 48* (1973), 322–350.

109. William Derham, *Physico-Theology: or a demonstration of the being and attributes of God, from his works of creation,* 3rd ed. (London, 1714), 169.

110. Charles de Linné, *Système de la nature,* trans. Vanderstegen de Putte from the 13th Latin edition of J. F. Gmelin (Bruselles, 1793), 14.

111. Linné, "Économie de la Nature," in Richard Pulteney, *Revue générale des écrits de Linné,* trans. L. A. Millin de Grandmaison, 2 vols. (London, 1789), *2,* 217.

112. Ibid., *2,* 291.
113. Ibid., *2,* 287.
114. Ibid., *2,* 218.
115. Buffon, "Le lièvre," in *Oeuvres philosophiques de Buffon,* 363–365.
116. Buffon, "La Nature: Seconde Vue!" in *Oeuvres philosophiques de Buffon,* 38.
117. Charles Bonnet, *Contemplation de la nature,* new ed., 3 vols. (Hamburg, 1782), *1,* 284.
118. Linné, *Système de la nature,* 15.
119. Lamarck, "Forêt," *Botanique, 2* (1788), 519.
120. Ibid., 519.
121. Compare with Buffon, "De la terre végétale," *Théorie de la terre,* in *Oeuvres de Buffon,* ed. Lamouroux and Desmarest, 7 (Paris, 1825), 156–157.

Chapter Three
Eighteenth-Century Views of Organic Mutability

1. PZ, *1,* 265–266.
2. RCV, 121–122. See also HNASV, *1,* 180–181. Compare with Diderot, *Le rêve de d'Alembert,* where Mademoiselle de l'Espinasse reports d'Alembert as saying "Once I have seen inert matter pass to the sensitive state, nothing more can surprise me." In *Oeuvres philosophiques de Diderot,* ed. Paul Vernière (Paris, 1964), 303. *Le rêve de d'Alembert* was not published until 1830.
3. Linnaeus, *Fundamenta Botanica* (Amsterdam, 1736), no. 157, p. 18. Cited by Bentley Glass, "Heredity and variation in the eighteenth century concept of species," in Glass *et al., Forerunners of Darwin,* 145.
4. Adanson, "Examen de la question, si les espèces changent parmi les plantes; nouvelles expériences tentées à ce sujet," *Mémoires de l'Académie Royale des Sciences* (1769), 383.
5. The story of Linnaeus' thoughts on the plant he named *Peloria* was told in the eighteenth century by Duchesne fils [Antoine-Nicolas Duchesne], *Histoire naturelle des fraisiers* (Paris, 1766), part 2, "Remarques particulières," 49–59, and by Adanson, "Si les espèces changent," 386–393. The subject has been treated recently by Glass, "Heredity and variation in the eighteenth century concept of species," 144–172, and by James L. Larson, *Reason and experience: the representation of natural order in the work of Carl von Linné* (Berkeley, 1971), 99–104.
6. Cited in Larson, *Reason and experience,* 103.
7. James Marchant, "Observations sur la nature des plantes, *Mémoires de l'Académie Royale des Sciences* (1719), 59–66.
8. Larson, *Reason and Experience,* 104.
9. Ibid., 105–109.

10. Michel Adanson, *Familles des plantes* (Paris, 1763), cxiii (marginal heading).

11. Ibid., cxii.

12. Ibid., cxiii.

13. Ibid.

14. Ibid., cxiv.

15. Duchesne discussed the origins of his work in a letter to "la Société des Naturalistes residents à Paris," dated 8 frimaire an 3 [28 November 1794]. Muséum National d'Histoire Naturelle, MS 298.

16. Duchesne, *Histoire naturelle des fraisiers,* part 2, 18–19.

17. Ibid., 19.

18. Ibid., 62–63.

19. Ibid., 63.

20. Ibid., part 1, 220–221. Duchesne constructed a diagram illustrating what he presumed to be the genealogical relationships of the ten races of strawberries with which he was familiar (part 1, 228).

21. Duchesne, "Sur les rapports entre les êtres naturels," *Magasin encyclopédique, 1* (1795), no. 6, 289–294.

22. Adanson, "Si les espèces changent."

23. Ibid., 418. The italics are Adanson's.

24. Ibid., 419.

25. On Kolreuter see Olby, *Origins of Mendelism,* 20–35.

26. Duchesne, *Histoire naturelle des fraisiers,* part 2, 63; Adanson, *Familles des plantes,* clix-clx.

27. Buffon, "L'Asne," *Histoire naturelle, générale et particulière, 4* (1753), 377–391. *Oeuvres philosophiques de Buffon,* 356.

28. *Oeuvres philosophiques de Buffon,* 355.

29. Ibid., 380–382.

30. Ibid., 394–413.

31. Ibid., 395.

32. Ibid., 413.

33. Roger, *Les sciences de la vie,* 578–581; Paul Farber, "Buffon and the concept of species," *Journal of the History of Biology, 5* (1972), 259–284; Peter J. Bowler, "Bonnet and Buffon: theories of generation and the problem of species," *Journal of the History of Biology, 6* (1973), 259–281.

34. Bowler, "Bonnet and Buffon," 273.

35. Pierre Louis Moreau de Maupertuis, *Oeuvres* (Lyon, 1768; reprinted Hildesheim, 1965), *2,* 164.

36. Ibid., 184.

37. Ibid., 129.

38. Ibid., 164.

39. On Diderot see Lester G. Crocker, "Diderot and eighteenth-century French transformism," in Glass et al., *Forerunners of Darwin,* 114–143; and Roger, *Les sciences de la vie,* 585–682.

40. Arthur O. Lovejoy, *The great chain of being: a study of the history of an idea* (Cambridge, Mass., 1936), chapter nine.

41. Cited by A. de Quatrefages, *Charles Darwin et ses precurseurs français* (Paris, 1870), 37.

42. On Bonnet see Lovejoy, *The great chain of being,* chapter nine, and Bowler, "Bonnet and Buffon."

43. Lovejoy, *The great chain of being,* 285.

44. Charles Darwin, *On the origin of species,* facsimile of the first edition (Cambridge, Mass., 1966), 481.

45. For an exposition of Rouelle's geological theories, see Antoine-Nicolas Desmarest, "Rouelle," *Encyclopédie méthodique: géographie-physique, 1* (1794), 409–431. See also Rhoda Rappaport, "G.-F. Rouelle: an eighteenth-century chemist and teacher," *Chymia, 6* (1960), 68–101.

46. De Maillet, *Telliamed, ou entretiens d'un philosophe indien avec un missionaire françois sur la diminution de la mer, la formation de la terre, l'origine de l'homme, &c.,* 2 vols. (Amsterdam, 1748). On the popularity or notoriety of de Maillet's work in the eighteenth century see Desmarest, "Maillet," *Encyclopédie méthodique: géographie-physique, 1* (1794), 319–327; and Mornet, *Les sciences de la nature en France au XVIIIe siècle,* 248. For more on de Maillet's work and for an English translation of *Telliamed* see Albert Carozzi (ed. and trans.), *Telliamed . . .* (Urbana, Ill., 1968).

47. Cited by Desmarest, "Holback [sic]," *Encyclopédie méthodique: géographie-physique, 1* (1794), 249.

48. Buffon, *Histoire naturelle, 1* (Paris, 1749), 440. Compare with Lamarck, *Hydrogéologie,* 44–47.

49. Desmarest, "Rouelle," 419.

50. Desmarest, "Holback," 248–249. This passage is taken almost word for word from Holbach, "Préface du traducteur," in Jean-Gotlob Lehmann, *Traités de physique, d'histoire naturelle, de minéralogie et de métallurgie,* 3 vols. (Paris, 1759), *3,* xi-xii.

51. Lavoisier, "Observations générales, sur les couches modernes horizontales," *Mémoires de l'Académie Royale des Sciences* (1789 [1793]), 368–369.

52. J. L. Giraud-Soulavie, *Histoire naturelle de la France méridionale,* 8 vols. (1780–1784), *1* (1780), 349. Cited by Léon Aufrère, *De Thales à Davis: le relief et la sculpture de la terre . . . Soulavie et son secret* (Paris, 1952), 61.

53. Aufrère, *Soulavie,* 61.

54. Foucault, *The order of things,* 155.

55. RCV, 141.

56. *Flore, 1,* xv, xxxii, xc.

57. Ibid., 174.

58. Ibid., iv.

59. Ibid., xxvi.
60. "Espèce" (1786), 395.
61. Ibid.
62. *Flore, 1,* 141.
63. "Herborisations," *Botanique, 3* (1789), 116.
64. Ibid. and "Jardin de botanique," *Botanique, 3* (1789), 212.
65. "Froment," *Botanique, 2* (1788), 557–558.
66. Ibid., 558.
67. Tschoudi, "Transplantation," *Supplement à l'encyclopédie, 4* (1777), 966b–976b. This article was published separately as *De la transplantation, de la naturalisation et du perfectionnement des végétaux* (London and Paris, 1778). Lamarck's esteem for Tschoudi's writings is expressed in the letter from Lamarck to Pancoucke published by Frans A. Stafleu, "Lamarck: the birth of biology," *Taxon, 20* (1971), 406–407.
68. Tschoudi, "Transplantation," 969b.
69. L. Reynier, "De l'influence du CLIMAT sur la forme et la nature des végétaux," *Journal d'histoire naturelle, 2* (1792), 101–148.
70. Antoine-Nicolas Duchesne, "Sur le fraisier de Versailles," *Journal d'histoire naturelle, 2* (1792), 345.
71. Reynier, "Mémoire pour servir à l'histoire de la marchant variable," *Journal de physique, 30* (1787), 171–174; "Relatif à la formation des corps, par la simple aggrégation de la matière organisée," *Journal de physique, 31* (1787), 102–108; "Sur la crystallisation des êtres organisés," *Journal de physique, 33* (1788), 215–217.
72. J. C. Lamétherie, "Discours préliminaire," *Journal de physique, 44* (1794), 8.
73. Lamarck, "Philosophie botanique," *Journal d'histoire naturelle, 1* (1792), 10; "Sur une nouvelle espèce de Vantane," *Journal d'histoire naturelle, 1,* 144–145; "Sur une nouvelle espèce de Grassette," *Journal d'histoire naturelle, 1,* 334–335.
74. See Chapter Seven, note 52.
75. Cuvier and Geoffroy, "Histoire naturelle des orang-outangs," *Magasin encyclopédique, 1* (1795), no. 3, 452. Also *Journal de Physique, 46* (1798), 185–186.

Chapter Four
The Preoccupations of the New Professor

1. *Faits physiques, 2,* 213–214. SASV, 1–48. The comments that Lamarck made in 1797 on the subject of generation (MPHN, 270) do not constitute proof that Lamarck still believed in species fixity at that date.
2. Augustin-Pyramus de Candolle, *Mémoires et souvenirs de Augustin-Pyramus de Candolle,* 58 (see also 54).

3. The full title that Lamarck initially intended to use was *Mémoires présentant les bases d'une nouvelle théorie, physique et chimique, fondée sur la considération des molécules essentielles des composés, et sur celle des trois états principaux du feu dans la nature; servant en outre de developpement à l'ouvrage intitulé: réfutation de la théorie pneumatique.*

4. "Discours prononcé à la Société Philomathique, le 23 floréal an 5," 1–2.

5. "De l'influence de la lune sur l'atmosphère terrestre," *Journal de physique, 46* (1798), 428–435.

6. "Mémoire sur la matière du feu, considéré comme instrument chimique dans les analyses," *Journal de physique, 48* (1799), 345–361; "Mémoire sur la matière du son," *Journal de physique, 49* (1799), 397–412; *Annuaire météorologique pour l'an VIII* (Paris, 1799); on Lamarck's geological memoir see Chapter One, note 3.

7. "Sur les genres de la Seiche, du Calmar et du Poulpe vulgairement nommés polypes de mer. Lu à la l'Institut National, le 21 floréal an 6 [10 May 1798]," *Mémoires de la Société d'Histoire Naturelle de Paris, 1* (1799), 1–25; "Prodrome d'une nouvelle classification des coquilles, comprenant une rédaction appropriée des caractères génériques, et l'établissement d'un grand nombre de genres nouveaux," *Mémoires de la Société d'Histoire Naturelle de Paris, 1* (1799), 63–91.

8. *Faits physiques, 1,* xii.

9. Ibid., 10.

10. Ibid., *2,* 26n.

11. "Apperçu analytique des connoissances humaines," *Inédits,* 70–82.

12. Lamarck identified these four considerations himself in his *Réfutation de la théorie pneumatique,* 9–13.

13. MPHN, 131.

14. Ibid., 131–237. See especially the table facing p. 227.

15. *Réfutation de la théorie pneumatique,* 6.

16. For a more extended discussion of Lamarck's chemistry see Leslie J. Burlingame, "Lamarck's theory of transformism in the context of his views of nature, 1776–1809," doctoral thesis, Cornell University, 1973, pp. 126–153 (esp. 142–149).

17. MPHN, 91.

18. *Réfutation de la théorie pneumatique,* 10.

19. MPHN, 94.

20. "Discours prononcé à la Société Philomathique, le 23 floréal an 5," 2.

21. MPHN, 9.

22. *Réfutation de la théorie pneumatique,* 13.

23. *Faits physiques, 2,* 289.

24. Ibid., *2,* 286. See also p. 288.

25. Ibid., *1,* 11.

26. See, for example, Ibid., *1,* 16–17.

27. Gillispie, "The formation of Lamarck's evolutionary theory," 331, 329.

28. "Espèce" (1817), 441.

29. *Faits physiques, 1,* 16; MPHN, 238.

30. *Faits physiques, 2,* 185.

31. MPHN, 255.

32. Ibid.

33. SASV, 387. For Spallanzani's observations, see Spallanzani, *Opuscules de physique, animale et végétale,* tr. Jean Senebier, 3 vols. (Paris, 1787), *2,* 203–249.

34. MPHN, 300.

35. Ibid., 289–290.

36. Ibid., 282. In his *Philosophie zoologique* of 1809 Lamarck maintained that from 1796 on he had been telling the students in his course that *"cellular tissue* is the matrix from which all the organs of the body have been successively formed, and that the *movement of fluids* through this tissue is the means nature uses to gradually create and develop these organs from this tissue." PZ, *2,* 47. The italics are Lamarck's.

37. "Sur les observations et le but de toute météorologie statistique," *Annales de statisque, 4* (1802), 133.

38. See Lamarck's *Annuaire météorologique* for the years 1800–1810 and his "Météorologie," *Dictionnaire, 20* (1818), 451–477.

39. *Annuaire météorologique pour l'an XIV,* 131–132.

40. Ibid., 204.

41. *Annuaire météorologique pour l'an 1808,* 198.

42. "Sur les variations de l'état du ciel," *Journal de physique, 56* (1802), 119, 136–137. See also "Recherches sur la periodicité présumée des principales variations de l'atmosphère," *Journal de physique, 52* (1801), 299, 313–314.

43. "Météorologie," 457–467. The parallel between Lamarck's accounts of organic and atmospheric changes has been noted by Burlingame, "Lamarck's theory of transformism," 154–190.

44. *Annuaire météorologique pour l'an 1807,* 3; *Annuaire météorologique pour l'an 1808,* 2; "Météorologie," 453.

45. "Sur les variations de l'état du ciel," 138; *Annuaire météorologique pour l'an 1807,* 3–4; *Annuaire météorologique pour l'an 1808,* 1–6; "Météorologie," 473–477.

46. See Chapter 1, note 3.

47. Lassus, "Sciences physiques," *Magasin encyclopédique, 5* (1799), no. 1, 233–234.

48. MPHN, 346.

49. Ibid., 346–347.

50. Antoine Baumé, *Chymie expérimentale et raisonnée, 3,* 301–334.

51. Daubenton, "Sur les couches du globe de la terre," *Séances des écoles normales, 1,* 283. Daubenton's lecture was delivered at the École Normale on February 25, 1795.

52. Lamarck, "Prodrome d'une nouvelle classification des coquilles," 66. On Bruguière see Georges Cuvier, "Extrait d'une notice biographique sur Bruguières [sic], lue à la Société Philomathique dans sa séance générale du 30 janvier 1799," *Recueil des éloges historiques . . . par G. Cuvier,* new ed. (Paris, 1861), *3,* 357–372.

53. Bruguière died 3 October 1798, not in 1799 as is often reported. See G. A. Olivier, *Voyage dans l'Empire Ottoman, l'Égypte et la Perse,* 6 vols. (Paris, 1801–1807), *6,* 515–517.

54. Cited in Landrieu, *Lamarck,* 46.

55. John Gwyn Jeffrey, *British Conchology, 5* (1869), 190, and Ed. Lamy, "Les conchyliologistes Bruguière et Hwass," *Journal de conchyliologie, 74* (1930), 42–43, report that Lamarck and Bruguière met weekly at the house of the conchologist Hwass, who was secretary at the Danish embassy in Paris. Hwass, according to Bruguière (*Encyclopédie méthodique: vers, 1* [1789], 129) was one of the greatest authorities in conchology in all of Europe.

56. "Prodrome d'une nouvelle classification des coquilles," 66.

57. Bruguière, "Introduction," *Encyclopédie méthodique: vers, 1* (1789), iii.

58. Bruguière, "Cones," *Encyclopédie méthodique: vers, 1,* part 2 (1792), 601.

59. Ibid. Climatic change on the earth's surface, Bruguière maintained, has proceeded "in an uninterrupted succession ever since an epoch unknown until now, but in a sense different than that which the celebrated Buffon believed should be adopted, which is contradicted by observations that are as precise as they are numerous."

60. Ibid.

61. Bruguière, "Sur les mines de charbon des montagnes des Cévennes, et sur la double empreinte des fougères qu'on trouve dans leurs schistes," *Journal d'histoire naturelle, 1* (1792), 109–131.

62. Cuvier, "Bruguière," 365; Bruguière, "Sur les mines de charbon," 122–123.

63. Bruguière, "Sur les mines de charbon," 120.

64. Ibid., 122–123.

65. Lamarck, *Hydrogéologie,* 88.

66. Fontenelle, *Entretien sur la pluralité des mondes,* in *Oeuvres complètes* (Paris, 1790), *2,* 134.

67. "Discours d'ouverture . . . an XI," in Giard (ed.), *Discours d'ouverture,* 103.

68. *Hydrogéologie,* 67.

69. Desmarest, "Anecdotes de la nature et de l'histoire naturelle de la terre," *Encyclopédie méthodique: géographie-physique, 2* (1803), 561.

70. Faujas de Saint-Fond, *Histoire naturelle de la montagne de Maestricht,* 13–14.

71. Lavoisier, "Observations générales, sur les couches modernes horizontales," *Mémoires de l'Academie Royale des Sciences* (1789 [1793]), 355.

72. *Hydrogéologie,* 179.

73. Ibid., 88.

74. Ibid., 177.

75. Ibid., 179.

76. Ibid., 37.

77. Ibid., 266–267 n.

78. "Mémoire sur les fossiles des environs de Paris," *Annales du Muséum National d'Histoire Naturelle, 1* (1802), 301.

79. *Hydrogéologie,* 87.

80. "Mémoire sur les fossiles des environs de Paris," 300 n.

81. Ibid.

82. "Discours préliminaire," *Botanique, 1* (1783), xliii.

Chapter Five
Invertebrate Zoology and the Inspiration of Lamarck's Evolutionary Views

1. Étienne Geoffroy Saint-Hilaire, *Discours prononcés sur la tombe de M. le chevalier de Lamarck* (Paris, n.d.), 7–8.

2. Bruguière, "Conchyliologie," *Encyclopédie méthodique: vers, 1,* 545–548.

3. Muséum National d'Histoire Naturelle, MS 464. "Procès-verbaux de la Société d'Histoire Naturelle," (19 July 1793), 109.

4. See, for example, PZ, *1,* 57–58.

5. For a contemporary's view of the Muséum and its riches, see Gotthelf Fischer, *Das Nationalmuseum der Naturgeschichte zu Paris,* 2 vols. (Frankfurt, 1802). See also Cuvier, *Leçons d'anatomie comparée, 1* (1800), ii.

6. "Objets d'histoire naturelle recueillis en Hollande," *Décade philosophique, 5* (an 3 [1795]), 535–536.

7. "Lettre du Représentatif du Peuple français aux États-Généraux," *Décade philosophique, 5* (an 3 [1795]), 112–114.

8. "Cabinet du Stathouder à Paris," *Magasin encyclopédique, 1* (1795), no. 2, 419.

9. Cuvier, "Mémoire sur les espèces d'éléphans vivantes et fossiles," *Magasin encyclopédique, 2* (1796), no. 3, 440–445, and *Mémoires de l'Institut de France, 2* (1799), 1–22; E. Geoffroy and G. Cuvier, "Mémoire sur les orang-outangs," *Journal de physique, 46* (1798), 185–191; and Lamarck, "Sur les genres de la Seiche . . ." (see above, Chapter Four, note 7).

10. For a recent discussion of this expedition and many of its difficulties, see John Dunmore, *French explorers in the Pacific, 2* (Oxford, 1969), 9–40.

11. A. L. Jussieu, "Notice sur l'expédition à la Nouvelle-Hollande, enterprise pour des recherches de géographie et d'histoire naturelle," *Annales du Muséum d'Histoire Naturelle, 5* (1804), 1–11.

12. Cuvier, "Mémoire sur la structure interne et externe, et sur les affinités des animaux auxquels on a donné le nom de vers; lu à la société d'Histoire naturelle, le 21 floréal de l'an 3," *La décade philosophique, 5* (1795), 385–396.

13. Ibid., 396.

14. Ibid., 388, 391–394.

15. "Nouvelles littéraires," *Magasin encyclopédique, 2* (1796), no. 1, 285.

16. Lamarck, MPHN, 314 (first table); "Sur les genres de la Seiche . . . ," 3.

17. Cuvier, *Tableau élémentaire de l'histoire naturelle des animaux* (Paris, 1798), vii; *Leçons d'anatomie comparée, 1* (1800), xix.

18. Cuvier, *Leçons d'anatomie comparée, 1* (1800), xx.

19. Cuvier, "Nouveau système des animaux sans vertèbres," *Magasin encyclopédique, 6* (1801), no. 6, 387–388.

20. Ibid., 387–388.

21. Lamarck, PZ, *1,* 118.

22. Ibid., 121.

23. Bruguière, *Encyclopédie méthodique: vers,* 1, vii.

24. Lamarck, PZ, *1,* 122.

25. Ibid., 122–123.

26. Lamarck, *Extrait du cours de zoologie* (Paris, 1812), 2.

27. Duchesne, "Classification des habitans des eaux," *Magasin encyclopédique, 2* (1796), no. 2, 300–301.

28. Ibid., 301. Duchesne's article offers no support for Lamarck's claim concerning the establishment of the other invertebrate classes, but it does suggest that there may be some validity in Lamarck's claim that independently of Cuvier he identified the mollusks as a separate class and ranked them between the fish and the insects. Considering that Lamarck's initial expertise as an invertebrate zoologist was limited primarily to his conchological studies, it is not unreasonable that Lamarck's earliest attempt at reforming the classification of the invertebrates might have been the identification of a class of "mollusks."

29. Lamarck, MPHN, 314 (second table); "Prodrome d'une nouvelle classification des coquilles."

30. "Nouvelles littéraires," *Magasin encyclopédique, 2* (1796), no. 1, 285.

31. Lamarck, "Discours . . . de l'an six," *Inédits,* 160.

32. Compare *Inédits,* 154–155, with Bruguière, *Encyclopédie méthodique: vers, 1,* i.

33. Compare Lamarck, *Inédits,* 156–160, with G. A. Olivier, "Mémoire sur l'utilité de l'étude des insectes, relativement à l'agriculture et aux arts," *Journal d'histoire naturelle, 1* (1792), 33–56 (notably 36–54).

34. Lamarck, *Inédits,* 160.

35. Ibid.

36. Daubenton, "Avertissement," *Encyclopédie méthodique: histoire naturelle, 1* (1782), v.

37. Daubenton, "Sur les rapports que l'on a recherchés entre les corps bruts et les corps organisés," *Séances des écoles normales, 4,* 3–13.

38. Ibid., "Introduction," *1,* xv.

39. Ibid., *1,* iv.

40. Daubenton, "Programme," *Séances des écoles normales, 1,* 91.

41. Daubenton, "Sur la nomenclature méthodique de l'histoire naturelle," *Séances des écoles normales, 1,* 433.

42. Ibid., 434.

43. Lamarck, SASV, 17n.

44. Lamarck, *Inédits,* 160.

45. Ibid., 160–161.

46. Ibid., 154.

47. Ibid., 154. Bruguière, *Encyclopédie méthodique: vers, 1,* i.

48. It may be noted that in a decidedly nonevolutionary framework Linnaeus identified the "principal operations of the inhabitants of nature" as follows: "(1). To multiply the species, so they are sufficient for their functions. (2). To conserve the equilibrium between the species of animals and plants, so that the same proportion is perpetuated. (3). To preserve themselves from destruction, so that the order is maintained." Linné, *Système de la nature,* 15.

49. Lamarck, *Inédits,* 154.

50. Lamarck, "Prodrome d'une nouvelle classification des coquilles." The memoir was read December 11, 1798.

51. Ibid., 66–67.

52. See Chapter One, note 9.

53. Cuvier, "Mémoire sur les espèces d'éléphans vivantes et fossiles. Lu à la séance publique de l'institut national, le 15 germinal an 4." Muséum National d'Histoire Naturelle, MS 628. The manuscript of this memoir is not identical to either the "extract" of it that appeared in the *Magasin encyclopédique, 2* (1796), no. 3, 440–445 or to the published version in *Mémoires de l'Institut, 2* (1799), 1–22.

54. "Les espèces d'éléphans," Muséum National d'Histoire Naturelle, MS 628, 15.

55. Ibid., 15–16; *Magasin encyclopédique, 2* (1796), no. 3, 444–445.

56. Muséum National d'Histoire Naturelle, MS 628, 14–15.

57. Faujas de Saint-Fond, *Histoire de la montagne de Saint-Pierre de Maestricht,* 28. This work was published in sections. The "Discours préliminaire," from which the quote is taken, appeared in 1799.

58. Ibid., 124–125. This section of the book was not published until 1801 at the earliest.

59. Ibid., 126–136. By 1803 Faujas was able to list fifty-six species of fossil shells with known living analogs. See B. Faujas-St.-Fond, *Essai de géologie, ou mémoires pour servir à l'histoire naturelle du globe, 1* (Paris, 1803), 58–75.

60. Lamarck, "Prodrome d'une nouvelle classification des coquilles," 67.

61. Ibid., 64.

62. Lamarck, SASV, 407.

63. HNASV, *1,* 323.

64. SASV, 23.

65. PZ, *1,* 75.

66. Ibid., 99–100.

67. Ibid., 101.

68. Camille Limoges, *La sélection naturelle* (Paris, 1970), 38.

69. Bruguière, *Encyclopédie méthodique: vers, 1,* 33, 472.

70. Bruguière, "Description de deux coquilles, des genres de l'Oscabrion et de la Pourpre," *Journal d'histoire naturelle, 1* (1792), 22.

71. Bruguière, *Encyclopédie méthodique: vers, 1,* 472.

72. Ibid.

73. Ibid., 32.

74. Lamarck, PZ, *1,* 75–76.

75. Lamarck, "Sur une nouvelle espèce de Trigonie, et sur une nouvelle d'Huitre, découvertes dans le voyage du capitaine Baudin," *Annales du Muséum d'Histoire Naturelle, 4* (1804), 351–359. See Stephen Jay Gould, "Trigonia and the origin of species," *Journal of the History of Biology, 1* (1968), 41–56.

76. PZ, *1,* 77. See also SASV, 408, and "Sur une nouvelle espèce de Trigonie," 352–353.

77. PZ, *1,* 77–78. See also SASV, 408–409.

78. Cuvier, "Extrait d'un ouvrage sur les espèces de quadrupèdes dont on a trouvé les ossemens dans l'intérieur de la terre, adressé aux savans et aux amateurs des sciences," *Magasin encyclopédique, 7* (1801), no. 1, 64.

79. Lamarck, SASV, 408–409.

80. On Lacépède see Chapter Seven. Jean-Claude Delamétherie's views on the origin of living things are to be found in his *Considérations sur les êtres organisés: de la perfectibilité et de la dégénérescence des êtres organisés,* vol. 3 (Paris, 1860), and in his *Leçons de géologie données au Collège de France,* 3 vols. (Paris, 1816) (see especially vol. 1, 346–354). In

his *Leçons de géologie* Delamétherie argued that if one were to be truly philosophical about the question of the production of organized beings, one would have to admit "that a prime movement has been imparted to matter, and that all subsequent movements are a consequence of the first" (*1,* 348, 354). He did not, however, believe in the transmutation of species. He supposed that the different species were spontaneously generated through a process of crystallization as the waters that initially covered the globe gradually diminished. Pierre Denys de Montfort did not set his own views down systematically, but aspects of his thinking about the origin of living things are to be found in his *Histoire naturelle, générale et particulière des mollusques, animaux sans vertèbres et à sang blanc,* 5 vols. (Paris: an 10 [1802]–an 13 [1805]), esp. vol. I.

81. Faujas de Saint-Fond, *Histoire naturelle de la montagne de Saint-Pierre de Maestricht,* 35.

82. Lamarck, MPHN, 272.

83. Lamarck, SASV, 41. Lamarck's new views on the origin of life were alluded to briefly in his introductory lecture of 1800 and in his discussion of "amorphous polyps" in his SASV (pp. 12, 41, 391–393), and then expounded in their essential features in 1802 in his RCV (pp. 68–123). Much of the discussion of "direct" generation in PZ, *2,* 61–90 is taken from RCV. The comparable section in HNASV, *1,* 165–181 is rewritten but not basically different from the earlier presentations.

84. RCV, 103.

85. PZ, *1,* 285. See also SASV, 390, 397.

86. PZ, *1,* xvi.

87. MPHN, 279.

88. SASV, 291; RCV, 101–102.

89. SASV, 41, 358–359.

90. "Histoire naturelle des animaux invertebrés ou à sang blanc: discours préliminaire an VI à an VII," *Inédits,* 147. The same passage appears in RCV, 95.

91. RCV, 98.

92. Ibid.,101. See also SASV, 391–392.

93. RCV, 103.

94. Ibid., 103–104. See also SASV, 41, 391–392; and PZ, *2,* 83.

95. RCV, 101.

96. Ibid., 69–70. The italics are Lamarck's.

97. Ibid., 121–122. This is stated in much the same terms in HNASV, *1,* 180–181: "After the removal of this first difficulty that the spontaneous generations of the beginning of each kingdom offer to us . . . all the others relative to the composition of organization in the animals and the formation of the different special organs observed among them seem to me to vanish easily."

98. SASV, 11.

99. *Histoire naturelle des végétaux,* vols. 1 and 2 (Paris: an 11 [1803]), *2,* 257–258. Though this work was published in 1803, Lamarck may have been finished with his part of it (Mirbel was responsible for vols. 3–15) in 1801 or early in 1802, for he does not refer in it to either of the books he published in 1802, but he does refer to it in *Hydrogéologie* (pp. 109–110). The *Histoire naturelle des végétaux* does contain references to SASV. Most of the *Histoire naturelle des végétaux* consists of articles Lamarck wrote earlier for the botanical section of the *Encyclopédie méthodique.*

100. *Histoire naturelle des végétaux, 2,* 257.

101. SASV, 17.

102. RCV, 38.

103. SASV, vi.

104. Ibid., 14–15.

105. RCV, 7–8. The claim of 1800 is repeated on p. 62.

Chapter Six
Lamarck's Theory of Evolution

1. HNASV, *1,* iii-iv.

2. See Robert A. Nisbet, *Social change and history: aspects of the Western theory of development* (London, Oxford, New York, 1969), 139–158. Nisbet is following the lead of Frederick J. Teggart, *Theory of history* (New Haven, 1925).

3. Cited by Nisbet, *Social change and history,* 157.

4. Lamarck, RCV, 7–8.

5. PZ, *1,* 221.

6. HNASV, *1,* 133.

7. See Daudin, *Cuvier et Lamarck, 2,* 13–35.

8. "Discours prononcé a la Société Philomathique, le 23 floréal an 5," 4.

9. PZ, *1,* 39–52.

10. Ibid. See also HNASV, *1,* 342–382.

11. PZ, *1,* 107.

12. Ibid., *1,* 132.

13. Ibid., *1,* 133.

14. Ibid., *1,* 70. See Lacépède, Cuvier, and Lamarck, "Rapport des professeurs du Muséum sur les collections d'histoire naturelle rapportées d'Egypte . . . ," *Annales du Muséum, 1* (1802), 234–241.

15. Ibid., 236–237. Cuvier did not discuss the theoretical significance of the similarity between the ancient ibis and the modern bird in his "Mémoire sur l'ibis des anciens Égyptiens," *Annales du Muséum, 4* (1804), 116–135.

16. PZ, *1,* 70.

17. Ibid., *1,* 70–71.

18. "Espèce" (1817), 448–449.
19. PZ, *1,* 265–267.
20. J. S. Wilkie, "Buffon, Lamarck, and Darwin: the originality of Darwin's theory of evolution," in P. R. Bell (ed.), *Darwin's biological work: some aspects reconsidered* (Cambridge, 1959), 292.
21. George Gaylord Simpson, *This view of life* (New York, 1964), 48.
22. RCV, 70. See also PZ, *1,* 363.
23. PZ, *1,* 403. Virtually the same definition appeared in RCV, 71.
24. PZ, *1,* 404.
25. Ibid., *1,* 403.
26. RCV, 78. See also PZ, *2,* 305–310.
27. PZ, *2,* 4–5.
28. HNASV, *1,* 229n.
29. Ibid., *1,* 42.
30. Ibid., *2,* 454.
31. Ibid.
32. PZ, *2,* 11, 17.
33. HNASV, *1,* 42–43. See also RCV, 159–160.
34. PZ, *2,* 5–6.
35. Ibid., *2,* 15–18.
36. HNASV, *1,* 171–172. See also Lamarck, *Système analytique des connaissances positives de l'homme* (Paris, 1820), 115–116.
37. PZ, *1,* 285. See also HNASV, *1,* 411.
38. PZ, *2,* 76.
39. HNASV, *1,* 126, 179.
40. Ibid., *1,* 433.
41. Ibid., *2,* 442–443.
42. Ibid., *2,* 443.
43. Ibid., *2,* 443–444.
44. Ibid., *2,* 444.
45. RCV, 8–9. Quoted in PZ, *2,* 52–53.
46. HNASV, *1,* 182–183.
47. PZ, *2,* 310–314.
48. HNASV, *1,* 215. See also PZ, *2,* 113.
49. "Discours préliminaire pour le cours de l'an six," *Inédits,* 152–153.
50. PZ, *2,* 115–116.
51. Ibid., *2,* 127–168.
52. Ibid., *2,* 137–138.
53. Ibid., *2,* 147.
54. Ibid., *2,* 148.
55. Ibid., *2,* 156–157.
56. Ibid., *2,* 161.
57. *Extrait du cours de zoologie du Muséum d'Histoire Naturelle, sur les animaux sans vertèbres . . .* (Paris, 1812), 8–10. HNASV, *1,* 379–381.

58. PZ, *1,* 159–160.

59. Ibid., *1,* 163.

60. *Extrait du cours de zoologie,* p. 29.

61. HNASV, *1,* 362.

62. Ibid., *1,* 323–324. "Nature," 373.

63. Lamarck's tendency to suppose too readily that nature's processes could be comprehended easily has been commented on by J. S. Wilkie, "Buffon, Lamarck, and Darwin," 299–300.

64. MPHN, "Premier tableau" facing p. 314.

65. SASV, 35.

66. RCV, 24, 37.

67. "Discours d'ouverture . . . mai 1806," in A. Giard (ed.), *Discours d'ouverture des cours de zoologie donnés dans le Muséum d'Histoire Naturelle (An VIII, an X, an XI et* 1806) (Paris, 1907), 142, 147–148. Giard's collection of Lamarck's discourses also appears in *Bulletin scientifique de la France et de la Belgique, 40* (1907), 439–595.

68. PZ, *1,* 209.

69. *Extrait du cours de zoologie,* 9; HNASV, *1,* 381.

70. HNASV, *1,* 451–462.

71. PZ, *2,* 462.

72. Ibid., *2,* 87–88.

73. Ibid., *2,* 88–89.

74. *Extrait du cours de zoologie,* 82.

75. HNASV, *1,* 455–456. On the subject of the spontaneous generation of parasitic worms see John Farley, "The spontaneous generation controversy (1700–1860): the origin of parasitic worms," *Journal of the History of Biology, 5* (1972), 95–125.

76. M.-J.-C. Savigny, "Recherches anatomiques sur les Ascidies composées et sur les Ascidies simples," in Savigny, *Mémoires sur les animaux sans vertèbres: seconde partie* (Paris, 1816), 1–66, 83–132. Ch. Alex. Lesueur, "Mémoire sur l'organisation des Pyrosomes," *Bulletin des sciences* (1815), 70–74. A. Desmarest and Lesueur, "Note sur le Botrylle étoilé (Botryllus stellatus) Pall.," *Bulletin des sciences* (1815), 74–78. See Daudin, *Cuvier et Lamarck, 1,* 278–285.

77. HNASV, *1,* 451.

78. Ibid., *3,* 88–89.

79. "De la disposition qu'il faut donner à la distribution des animaux," *Inédits,* 277.

80. Ibid.

81. HNASV, *1,* 457.

82. "De la disposition qu'il faut donner à la distribution des animaux," *Inédits,* 279.

83. HNASV, *1,* 455.

84. Ibid., *1,* 454.

85. Ibid., *3,* 90.

86. PZ, *1,* 221.
87. SASV, 13.
88. Ibid.
89. Ibid., 14–15.
90. RCV, 7–9.
91. Ibid., 42.
92. Ibid., 39.
93. PZ, *1,* 235.
94. "Discours de l'an XI," in Giard (ed.), *Discours d'ouverture,* 89.
95. HNASV, *1,* 200. For other writers before Lamarck who believed in the inheritance of acquired characters see Conway Zirkle, "The early history of the idea of acquired characters and of pangenesis," *Transactions of the American Philosophical Society,* n.s. *35* (1946), 91–151.
96. HNASV, *1,* 191.
97. "Instinct," *Dictionnaire, 16* (1817), 340. See also PZ, *2,* 305 and HNASV, *1,* 251.
98. PZ, *2,* 310.
99. "Instinct," 338. HNASV, *1,* 245.
100. *Histoire naturelle des végétaux, 1,* 174, 233–234.
101. PZ, *1,* 223.
102. Ibid., *1,* 230. As indicated above, Chapter 3, note 69, the influence of habitat upon the appearance of *Ranunculus aquatilis* was discussed by Reynier in 1792.
103. PZ, *1,* 225.
104. "Habitude," *Dictionnaire, 14* (1817), p. 128. See also PZ, *2,* 307. Lamarck's thoughts on the mechanism of change in the plants and the simplest animals contradicts an often-cited passage in the *Philosophie zoologique* where he claimed: "whatever the circumstances be, they never bring about directly any modification whatsoever of the form and the organization of animals." (PZ, *1,* 221).
105. "Habitude," 129.
106. Ibid.
107. HNASV, *1,* 17, 242.
108. Ibid., *1,* 243–244.
109. PZ, *2,* 279.
110. "Instinct," 332. See also PZ, *2,* 255, 373.
111. HNASV, *1,* 17–18. See also "Instinct," 334.
112. HNASV, *3,* 238–239.
113. Ibid., *1,* 457.
114. Ibid., *3,* 268–269. See also "Habitude," 131.
115. HNASV, *3,* 269–270. See also PZ, *2,* 326, and "Habitude," 133.
116. PZ, *2,* 457–458.
117. SASV, 13–14.
118. See Pluche (Abbé Noël Antoine), *Le spectacle de la nature, ou entretiens sur les particularités de l'histoire naturelle qui ont paru les*

plus propres à rendre les jeunes gens curieux & à leur former l'esprit (Paris, 1782), *1,* 294. One of Pluche's interlocuters remarks of birds that "the beak, the claws, the length of the wings, and generally all the parts of their bodies have been arranged according to their needs. These are instruments adjusted to the nature of their activities and their way of life."

119. PZ, *1,* 240–248.

120. Ibid., *1,* 256–257.

121. Ibid., *1,* 252.

122. Ibid., *1,* 256.

123. This point has been stressed by Limoges, *La sélection naturelle,* 40–42.

124. SASV, 14.

125. PZ, *2,* 460–461.

126. Ibid., *1,* 349. See also RCV, 134.

127. RCV, 124.

128. PZ, *1,* 352.

129. Ibid., *1,* 351.

130. See for example RCV, 16–20, and PZ, *2,* 456–461. J. Schiller, "L'échelle des êtres et la série chez Lamarck," in Schiller (ed.), *Colloque international "Lamarck,"* 94–97, stresses that Lamarck's classes were identified according to distinct plans of organization, and that *discontinuity* therefore characterized Lamarck's view of the animal series. According to Schiller, "The progression that the organization of the classes represents does not imply the existence of intermediate forms representing a quantitative mean between two other types of organization. The change is abrupt and corresponds to the notion of mutation in the modern sense of the word." (p. 96). It was Lamarck's claim, however, that "when the circumstances require it, nature passes from one system to the other, without making any jump [*sans faire de saut*], provided [the systems] are adjacent." RCV, 43.

131. HNASV, *1,* 185.

132. Ibid., *1,* 185–186.

133. Ibid., *1,* 188.

134. Ibid.

135. Cuvier, "Éloge de Lamarck," xx.

136. Darwin's copy of the second edition of Lamarck's *Histoire naturelle des animaux sans vertèbres* is at Cambridge University Library. The annotation appears on page 157 of volume 1 (1835).

137. RCV, 61.

138. PZ, *1,* 64.

139. "Espèce" (1817), 446.

140. Ibid., 450.

141. Lacépède, *Histoire naturelle des poissons, 2,* xxxix–xli; *3,* liii.

142. PZ, *1,* 235.

143. RCV, 53–54.
144. HNASV, *1,* 200.
145. Ibid., *1,* 201.
146. PZ, *1,* 261–262.
147. See, for example, PZ, *1,* 60–63.
148. PZ, *1,* 226.
149. Ibid., *1,* 227.
150. Ibid., *1,* 240–241.
151. Ibid., *2,* 461.
152. "Espèce" (1817), 447–448.
153. Ibid., 448.
154. Ibid., 448–449.
155. Ibid., 448.
156. Ibid., 447. PZ, *1,* 59.
157. "Espèce" (1817), 447.
158. Ibid., 443.
159. Ibid., 445.
160. Ibid., 443.
161. Ibid., 443–444.
162. "Discours de l'an XI," in Giard (ed.), *Discours d'ouverture,* 100.
163. Ibid., 100–101.

Chapter Seven
The Frustrations and Consolations of the Naturalist-Philosopher

1. PZ, *2,* 450.
2. L. A. G. Bosc, *Histoire naturelle des coquilles, contenant leur description, les moeurs des animaux qui les habitent et leurs usages,* 5 vols. (Paris, an 10 [1802]), *1,* 64.
3. Ibid., *1,* 65.
4. Ibid., *2,* 149–150.
5. Muséum National d'Histoire Naturelle, MS 299, "Relevé des Mémoires lus et déposés à la Société d'Histoire Naturelle depuis sa formation jusqu'à sa Séance du —— [sic] inclusivement."
6. Bosc, *Histoire naturelle des coquilles, 3,* 240.
7. F. J. Gall and J. C. Spurzheim, *Des dispositions innés de l'âme et de l'esprit, du matérialisme, du fatalisme et de la liberté morale, avec des réflexions sur l'éducation et sur la législation criminelle* (Paris, 1811), 100.
8. De Bonald, *Recherches philosophiques sur les premiers objects des connoissances morales,* 2 vols. (Paris, 1818), *2,* 211–214.
9. Pierre-André Latreille, *Cours d'entomologie, ou de l'histoire naturelle des crustacés, des arachnides, des myriapodes et des insectes* (Paris, 1831), 21.

10. William Kirby and William Spence, *An introduction to entomology, or elements of the natural history of insects,* 4 vols. (London, 1818–1826), *3,* 350n.

11. Ibid., *3,* 348.

12. Ibid., *3,* 350.

13. William Kirby, *On the power, wisdom and goodness of God, as manifested in the creation of animals, and in their history, habits and instincts* (Philadelphia, 1837), xxiv.

14. Ibid., xxvi–xxvii.

15. Ibid., xxvii.

16. Ibid.

17. Ibid., xxviii–xxix.

18. Gall and Spurzheim, *Des dispositions innés de l'âme,* 97–98.

19. Tourlet, review of Lamarck, *Philosophie zoologique,* in *Gazette nationale ou le Moniteur Universel,* 18 October 1809, 1154.

20. Tourlet, review of Lamarck, *Recherches sur l'organisation des corps vivans,* in *Gazette nationale ou le Moniteur Universal,* 28 fructidor an 10 (15 Sept. 1802), 1462.

21. Ibid.

22. "Discours prononcé par le citoyen Cuvier à l'ouverture du cours d'anatomie comparée qu'il fait au Muséum national d'histoire naturelle, pour le citoyen Mertrud," *Magasin encyclopédique, 1* (1795), no. 5, 149.

23. "Mémoire sur les espèces d'éléphans vivantes et fossiles," *Magasin encyclopédique, 2* (1796), no. 3, 444–445.

24. *Recherches sur les ossemens fossiles de quadrupèdes,* 3.

25. William Coleman, *Georges Cuvier, zoologist: a study in the history of evolution theory* (Cambridge, Mass., 1964). On Cuvier see also "Georges Cuvier: journée d'études organisées par l'Institut d'Histoire des Sciences de l'Université de Paris," *Revue d'histoire des sciences, 23* (1970), 7–92.

26. Institut de France, Fonds Cuvier, MS 3136, "Cours de l'Atheneé de l'an XIII: discours préliminaire," 205.

27. *Leçons d'anatomie comparée* (Paris, an 8 [1800]), *1,* 47.

28. Muséum National d'Histoire Naturelle, MS 634, "Examen du Squélette d'un très grand Quadrupède, trouvé sous terre au Paraguay, et déposé au cabinet de Madrid, dont les figures ont été remises à la classe par le C.[n] Grégoire." This was the creature Cuvier decided to call *Megatherium fossile.*

29. What Cuvier called "conditions of existence" were what others called "final causes." See Cuvier, *Le Règne animal distribué d'après son organisation* (Paris, 1817), *1,* 6.

30. "Mémoire sur les espècces d'éléphans vivantes et fossiles," *Mémoires de l'Institut* (*class. math. phys.*), *2* (1799), 12. Italics added.

31. *Recherches sur les ossemens fossiles de quadrupèdes,* 74.

32. Ibid., 75–79.
33. Ibid., 76–79.
34. Ibid., 79–80.
35. Cited in John Viénot, *Le Napoléon de l'intelligence: Georges Cuvier* (Paris, 1932), 100.
36. "Rapport de l'Institut national . . . sur un ouvrage de M. André, ayant pour titre: théorie de la surface actuelle de la terre," *Journal des mines, 21* (1807), 421.
37. "Éloge de Lamarck," ii.
38. Ibid., xx.
39. See Coleman, *Georges Cuvier,* 161–164.
40. Institut de France, Fonds Cuvier, MS 3065, "Fragments et premières pages de l'introduction à l'ouvrage sur la variété de composition des animaux préparé par Mr Cuvier," 123.
41. *Histoire des sciences naturelles, depuis leur origine jusqu'à nos jours, chez tous les peuples connus,* ed. M. Magdeleine de Saint-Agy (Paris, 1841), *3,* 88.
42. "Éloge de Lamarck," xx–xxi.
43. Lamarck, PZ, *1,* 149 n.
44. Cuvier, *Leçons d'anatomie comparée,* 2nd ed. (Paris, 1835), *1,* 101.
45. Ibid.
46. Institut de France, Fonds Cuvier, MS 3065, "Sur la variété de composition des animaux," 109.
47. "Éloge de Lamarck," xx.
48. Étienne Geoffroy Saint-Hilaire, *Fragments biographiques, précédés d'études sur la vie, les ouvrages et les doctrines de Buffon* (Paris, 1838), 81.
49. *Leçons d'anatomie comparée,* 2nd ed., *1,* 101.
50. Frédéric Cuvier, "Observations préliminaires," in Georges Cuvier, *Recherches sur les ossemens fossiles,* 4th ed. (Paris, 1834), *1,* viii.
51. See in the published works Cuvier's *Histoire des sciences naturelles, 3,* 85–88; *Leçons d'anatomie comparée,* 2nd ed., *1,* 99–102; and the general comments in the "Éloge de Lamarck."
52. *Recherches sur les ossemens fossiles de quadrupèdes,* 1st ed., 1, 28. The footnote reads: "Voyez la *Physique de Rodig,* p. 106. Leipsig, 1801; et la p. 169 du 2e tome de Telliamed. M. de Lamarck est celui qui a developpé dans ces derniers temps ce système avec le plus de suite et la sagacité la plus soutenue dans son *Hydrogéologie* et dans sa *Philosophie zoologique.*" In the 1830 edition of the *Discours sur les revolutions . . . du globe* and in the fourth edition of the *Ossemens fossiles* the phrase "et la sagacité la plus soutenue" was dropped from the footnote. Cuvier's reference to Telliamed is apparently to the 1749 edition of that work, where, beginning on page 169, volume 2, the idea that flying fish may be transformed into birds is presented. The Rodig work referred to has the title

Lebende Natur. Page 106 of Rodig's work also has a discussion of the transformation of fish into birds.

53. *Recherches sur les ossemens fossiles,* 1st ed., *1,* 28.

54. Muséum National d'Histoire Naturelle, MS 631, 35–36. The original is as follows: "que l'habitude de mâcher par example, finit au bout de quelques siècles par leur donner des dents; l'habitude de marche [sic], leur donna des jambes; les canards à force de plonger devinrent des brochets; les brochets à force de se trouver à sec se changèrent en canards; les poules en cherchant leur pature au bord des eaux, et en s'efforçant de ne pas se mouiller les cuisses, reussirent si bien à s'alonger les jambes qu'elles devinrent des hérons ou des cigognes. Ainsi se formèrent par dégrés ces cent milles races diverses, dont la classification embarrasse si cruellement la race malheureuse que l'habitude a changée en naturalistes." It may be noted, as Coleman (*Georges Cuvier,* 191) has done, that the manuscripts of Cuvier's published works almost invariably correspond precisely to the published works themselves. An omission of the sort represented by this passage is quite rare.

55. See the first two items cited in note 51 and Cuvier's article "Nature," *Dictionnaire des sciences naturelles, 34* (1825), 261–268. In the manuscript cited in the previous footnote Cuvier attributes the views in question to some writers "in Germany."

56. On Lacépède see Roger Hahn, "Sur les debuts de la carrière scientifique de Lacépède," *Revue d'histoire des sciences, 27* (1974), 347–353, and "L'autobiographie de Lacépède retrouvée," *Dix-huitième siècle, 7* (1975), 49–85, as well as Cuvier, "Éloge historique de Lacépède," *Recueil des éloges historiques, 2,* 371–409.

57. Institut de France, Fonds Cuvier, MS 3222 (27). In a letter dated 15 messidor an 8 [4 July 1800] Lacépède thanks Cuvier for Cuvier's generous comments on Lacépède's book. Cuvier's observations on the book are recorded briefly in "Notice de la classe des sciences mathématiques et physiques de l'Institut National . . . ," *Magasin Encyclopédique, 6* (1800), no. 2, 382–383.

58. Viénot, *Le Napoléon de l'intelligence,* 100.

59. Lamarck, PZ, *1,* 241–242.

60. Cited in Coleman, *Cuvier,* 154.

61. Geoffroy Saint-Hilaire, *Fragments biographiques,* 81.

62. Lamarck, *Annuaire météorologique pour l'an 1810* (Paris, 1809), 167.

63. The only example Lamarck presented in HNASV was that of the formation of gastropod tentacles (*1,* 188). See Chapter Six.

64. The desire of some naturalists not to offend Cuvier is well illustrated by two letters in the Fonds Cuvier at the Institut de France (MS 3060, nos. 1 and 2) that were written to Cuvier in 1830 after Etienne Geoffroy Saint-Hilaire used the occasion of an academy report to raise

the issue of the unity of composition. The report, ostensibly by Latreille as well as Geoffroy, dealt with a memoir by Laurencet and Meyranx on the organization of the cephalopod molluscs. Both Latreille and Meyranx wrote to Cuvier disavowing any sympathy with Geoffroy's views and indicating their great desire to stay in Cuvier's good graces.

65. F.-V. Raspail, "Monstruosités remarquables," *Annales des sciences d'observation, 3* (1830), 277. The title of the article refers to teratological studies by Étienne Geoffroy Saint-Hilaire. Lamarck also characterized ridicule as a "deadly weapon." See his *Annuaire météorologique pour l'an 1807,* 4.

66. Cuvier and Geoffroy, "Histoire naturelle des Orang-Outangs," *Magasin encyclopédique, 1* (1795), no. 3, 452; *Journal de physique, 46* (1798), 185–186. See also F. Bourdier, "Lamarck et Geoffroy Saint-Hilaire face au problème de l'évolution biologique," *Revue d'histoire des sciences, 25* (1972), 311–325.

67. Institut de France, Fonds Cuvier, MS 3092, "Cours de College de France, 1830–1831. XVIIIe siècle. 1re partie," 37.

68. See Chapter 5.

69. See for example Faujas de Saint-Fond, *Essai de géologie, 1,* 153.

70. Lacépède, "Sur la durée des espèces," *Histoire naturelle des poissons,* 5 vols. (Paris, 1798–1802), *2* (1800), xxiii–lxiv.

71. Ibid., xxii–xxxiii.

72. Ibid., xxxiv.

73. Ibid.

74. Ibid., xxxv.

75. Ibid.

76. Ibid., xxxvii–xxxix.

77. Ibid., xl–xli.

78. Ibid., xlviii–xlix.

79. Ibid.

80. Ibid., lvi–lvii.

81. Ibid., lviii.

82. Ibid., lx.

83. Ibid., lxiii–lxix.

84. "Des effets de l'art de l'homme sur la nature des poissons," *Histoire naturelle des poissons, 3* (an 10 [1802]), i–lxvi.

85. Ibid., lxi–lxii.

86. Ibid., liii.

87. Ibid., lxiii.

88. Ibid., lxiii–lxiv.

89. Cuvier, "Éloge historique de Lacépède," *2,* 409.

90. On Geoffroy see Isidore Geoffroy Saint-Hilaire, *Vie, travaux, et doctrine scientifique d'Étienne Geoffroy Saint-Hilaire* (Paris, 1847); Théophile Cahn, *La vie et l'œuvre d'Étienne Geoffroy Saint-Hilaire*

(Paris, 1962); and F. Bourdier, "Lamarck et Geoffroy Saint-Hilaire" (note 66 above).

91. Recherches sur l'organisation des Gavials . . . ," *Mémoires du Muséum National d'Histoire Naturelle, 12* (1825), 153.

92. "Rapport fait à l'Académie des Sciences sur un mémoire de M. Roulin, ayant pour titre: sur quelques changemens observés dans les animaux domestiques transportés de l'ancien monde dans le nouveau continent," *Mémoires du Muséum National d'Histoire Naturelle, 17* (1828), 201–208 (see 206).

93. "Recherches sur l'organisation des Gavials," 153.

94. "Sur le degré de l'influence du monde ambiant pour modifier les formes animales; question intéressant l'origine des espèces téléosauriennes et successivement celle des animaux de l'époque actuelle," *Mémoires de l'Académie Royale des Sciences,* 2nd ser. *12* (1833), 63–92.

95. Ibid., 67.

96. Ibid., 69.

97. Ibid., 68.

98. Ibid., 77.

99. Ibid., 80–81.

100. Ibid., 81.

101. Ibid., 89n.

102. Ibid., 89–90n.

103. Lacépède, "Discours sur la pêche . . . ," *Histoire naturelle des poissons, 5* (an 11 [1803]), lvxiii. See also Hahn, "L'autobiographie de Lacépède retrouvée," 74. Lacépède continued to be active as a scientific functionary, however, and did publish some more articles together with his *Histoire naturelle des cétacés.*

104. Faujas de Saint-Fond, *Essai de géologie, 1,* 23.

105. RCV, iv.

106. Lamarck, "La biologie," presented by Pierre-P. Grassé, *Revue scientifique, 5* (1944), 267–276.

107. *Système analytique des connaissances positives de l'homme* (Paris, 1820). See also Lamarck's "Apperçu analytique des connoissances humaines," *Inédits,* 51–141.

108. Cuvier, Institut de France, Fonds Cuvier, MS 3092, "Cours du Collège de France, 1830–31. XVIIIe siècle. 1re partie," 36. On Geoffroy, see note 92 above.

109. G. P. Deshayes, *Description des animaux sans vertèbres découverts dans le bassin de Paris* (Paris, 1860), *1,* 35–36.

110. Desmarest, Lacépède, Latreille, Olivier, and Péron all paid attention to the geographical distribution of organisms.

111. "Espèce" (1817), 448.

112. See, for example, the reviews of the *Vestiges* by David Brewster in *North British Review, 3* (1845), esp. 500–501; Asa Gray in *North*

American Review, 62 (1846), esp. 490–494; and Adam Sedgwick in *Edinburgh Review* (1845), esp. 7. All these reviewers, while denying the validity of Lamarck's hypothesis, confess the superiority of his theory over that of the author of the *Vestiges.*

113. Lamarck complained in "Météorologie," *Dictionnaire, 20* (1818), 476, about the way his meteorological studies had been thwarted and the way current meteorological theory had been misrepresented (by Cuvier).

114. Muséum National d'Histoire Naturelle, MS 756, "Physique terrestre," 1st cahier, 7.

115. Ibid., 7–8.

116. Ibid., 9–10.

117. "L'homme," *Dictionnaire, 15* (1817), 270.

118. Ibid., 270–271n.

119. Landrieu, *Lamarck,* 63.

120. Ibid., 80.

121. "Nature," 398. Lamarck did not agree with Rousseau that happiness was incompatible with the development of the sciences. The problem as Lamarck saw it was that knowledge was not widely enough distributed (p. 397).

122. Ibid., 398.

123. Ibid.

124. Ibid., 399.

125. HNASV, *1,* 329–330n.

126. Ibid., 330n.

127. Ibid., 329.

128. *Système analytique des connaissances positives de l'homme,* 118.

129. Ibid.

130. HNASV, *1,* 333.

131. Ibid., 331.

132. MPHN, 255.

Bibliography

MANUSCRIPTS

The most important collection of Lamarck's manuscript writings is located at the Muséum d'Histoire Naturelle in Paris. This collection is described by Max Vachon, "À propos des manuscrits de Lamarck conservés à la bibliothèque centrale du Muséum National d'Histoire Naturelle de Paris (note préliminaire)," *Bulletin du Muséum National d'Histoire Naturelle, 39* (1967), 1023–1027, and by Max Vachon, Georges Rousseau, and Yves Laissus, "Liste complète des manuscrits de Lamarck conservés à la bibliothèque centrale du Muséum national d'histoire naturelle de Paris," *Bulletin du Muséum National d'Histoire Naturelle, 40* (1968), 1093–1102. A significant portion of these manuscripts has recently been published: Max Vachon, Georges Rousseau, and Yves Laissus (eds.), *Inédits de Lamarck, d'après les manuscrits conservés à la bibliothèque centrale du Muséum National d'Histoire Naturelle de Paris* (Paris, 1972).

Remaining unpublished among the Lamarck manuscripts at the Muséum d'Histoire Naturelle are Lamarck's interesting introduction to his *Physique Terrestre* (portions of which are quoted in the present study), some writings on meteorology and chemistry, some descriptive botany and zoology, and some notes taken from the writings of other authors. Not included in the *Inédits* volume, but published previously, is the short manuscript entitled "La biologie," in Pierre-Paul Grassé, "La biologie, texte inédit de Lamarck," *Revue scientifique, 5* (1944), 267–276. The only other important collection of Lamarck manuscripts known to exist is owned by Harvard University: these have been published in their entirety in William Morton Wheeler and Thomas Barbour (eds.), *The Lamarck manuscripts at Harvard* (Cambridge, Mass., 1933).

In addition to working with the Lamarck manuscripts at the Muséum d'Histoire Naturelle, I consulted the Cuvier papers at the Muséum d'Histoire Naturelle and at the Institut de France. The Cuvier papers at the Institut de France are very well catalogued (see H. Dehérain, *Catalogue des manuscrits du fonds Cuvier* [Paris, 1908]). The Muséum's Cuvier manuscripts, by contrast, are described only briefly (see A. Boinet, *Catalogue général des manuscrits des bibliothèques publiques de France: Paris, tome II, Muséum d'Histoire Naturelle* [Paris, 1914]). I also consulted the manuscript records of the Société Linnéenne de Paris at the Bibliothéque Mazarine (MS 4441) and of the Société d'Histoire Naturelle de Paris at the Muséum d'Histoire Naturelle (MS 298–299). Roger Hahn kindly lent me his notes on the *procès-verbaux* of the Société d'Histoire Naturelle (Muséum d'Histoire Naturelle MS 464).

PUBLISHED WORKS

Lamarck

An all but exhaustive bibliography of Lamarck's publications is to be found in Marcel Landrieu, *Lamarck, le fondateur du transformisme* (Paris, 1909), pp. 448–470. I cite here in chronological order those of Lamarck's writings published in his own lifetime that have been important for my analysis of Lamarck's thought, including four articles from the *Annales de statistique* of 1802 that were unknown to Landrieu.

1778 [1779] *Flore françoise, ou description succincte de toutes les plantes qui croissent naturellement en France, disposée selon une nouvelle méthode d'analyse et à laquelle on a joint la citation de leurs vertus les moins équivoques en médecine, et de leur utilité dans les arts.* 3 vols. Paris.

1783–1789 [1792] *Encyclopédie méthodique: botanique.* Vols. 1–3. Paris. The separate sections of Lamarck's contribution to the *Encyclopédie méthodique: botanique* have been dated as follows: *1,* 1–344 (1783), 345–752 (1785); *2,* 1–400 (1786), 401–774 (1788); *3,* 1–360 (1789), 361–760 (1792).

1784 "Mémoire sur un nouveau genre de plante nommé Brucea et sur le faux Bresillet d'Amérique," *Mémoires de l'Académie Royale des Sciences* (1784), 342–347.

1785 "Mémoire sur les classes les plus convenables à établir parmi les végétaux, et sur l'analogie de leur nombre avec celles déterminées dans le règne animal, ayant égard de part et d'autre à la perfection graduée des organes," *Mémoires de l'Académie Royale des Sciences* (1785), 437–453.

1788 "Mémoire sur le genre du Muscadier, Myristica," *Mémoires de l'Académie Royale des Sciences* (1788), 148–168.

1789 *Considérations en faveur du Chevalier de la Marck, ancien officier au régiment de Beajolais, de l'Académie royale des Sciences, botaniste du Roi, attaché au cabinet d'histoire naturelle.* Paris. Reproduced in full in Landrieu, *Lamarck,* pp. 34–36.

1789 *Mémoire sur le projet du Comité des finances relatif à la suppression de la place de botaniste, attaché au cabinet d'histoire naturelle.* Paris. Reproduced in full in Landrieu, *Lamarck,* pp. 36–39.

1790 (?) *Mémoire sur les cabinets d'histoire naturelle et particulièrement sur celui du jardin des plantes; contenant l'exposition du régime et de l'ordre qui conviennent à cet établissement pour qu'il soit vraiment utile.*

1791 "Instructions aux voyageurs autour du monde sur les observations les plus essentielles à faire en botanique," *Bulletin des sciences par la Société Philomathique de Paris, 1,* 8.

1792 "Sur l'histoire naturelle en général," *Journal d'histoire naturelle, 1,* 3–6.

1792 "Sur la nature des articles de ce journal qui concernent la botanique," *Journal d'histoire naturelle, 1,* 7–9.

1792 "Philosophie botanique," *Journal d'histoire naturelle, 1,* 9–19.

1792 "Sur le Calodendrum," *Journal d'histoire naturelle, 1,* 56–62.

1792 "Philosophie botanique," *Journal d'histoire naturelle, 1,* 81–92.

1792 "Sur les travaux de Linnaeus," *Journal d'histoire naturelle, 1,* 136–144.

1792 "Sur une nouvelle espèce de Vantane," *Journal d'histoire naturelle, 1,* 144–148.

1792 "Sur les systèmes et les méthodes de botanique, et sur l'analyse," *Journal d'histoire naturelle, 1,* 300–307.

1792 "Sur une nouvelle espèce de Grassette," *Journal d'histoire naturelle, 1,* 334–338.

1792 "Sur l'étude des rapports naturels," *Journal d'histoire naturelle, 1,* 361–371.

1792 "Sur les relations dans leur port ou leur aspect, que les plantes de certaines contrées ont entr'elles, et sur une nouvelle espèce d'Hydrophylle," *Journal d'histoire naturelle, 1,* 371–376.

1792 "Sur l'augmentation continuelle de nos connoissances à l'égard des espèces, et sur une nouvelle espèce de Sauge," *Journal d'histoire naturelle, 2,* 41–47.

1792 "Sur l'augmentation remarquable des espèces, dans beaucoup de genres qui n'en offroient depuis longtemps qu'une seule, et particulièrement sur une nouvelle espèce d'Helenium," *Journal d'histoire naturelle, 2,* 210–215.

1792 "Observations sur les coquilles, et sur quelques'uns des genres qu'on a établis dans l'ordre des vers testacés," *Journal d'histoire naturelle, 2,* 269–280.

1792 "Sur quatre espèces d'Hélices," *Journal d'histoire naturelle, 2,* 347–353.

1792 "Sur les ouvrages généraux en histoire naturelle; et particulièrement sur l'édition du Systema Naturae de Linneus, que M. J. F. Gmelin vient de publier," *Actes de la Société d'Histoire Naturelle de Paris, 1,* 81–85.

1794 *Recherches sur les causes des principaux faits physiques, et particulièrement sur celles de la combustion, de l'élévation de l'eau dans l'état de vapeurs; de la chaleur produite par le frottement des corps solides entre eux; de la chaleur qui se rend sensible dans les décompositions subites, dans les effervescences et dans le corps de beaucoup d'animaux pendant la durée de leur vie; de la causticité, de la saveur et de l'odeur de certains composés; de la couleur des corps; de l'origine des composés et de tous les minéraux; enfin de l'entretien de la vie des êtres organiques, de leur accroissement, de leur état de vigueur, de leur dépérissement et de leur mort.* 2 vols. Paris.

1796 *Réfutation de la théorie pneumatique ou de la nouvelle doctrine des chimistes modernes, présentée, article par article, dans une suite de réponses aux principes rassemblés et publiés par le citoyen Fourcroy dans sa Philosophie chimique; précédée d'un supplément complémentaire de la théorie exposée dans l'ouvrage intitulée: Recherches sur les causes des principaux faits physiques, auquel celui-ci fait suite et devient nécessaire.* Paris.

1797 *Mémoires de physique et d'histoire naturelle, établis sur des bases de raisonnement indépendantes de toute théorie; avec l'exposition de nouvelles considérations sur la cause générale des dissolutions; sur la matière du feu; sur la couleur des corps; sur la formation des composés; sur l'origine des minéraux, et sur l'organisation des corps vivans, lus à la première classe de l'Institut National, dans ses séances ordinaires.* Paris.

1798 "De l'influence de la lune sur l'atmosphère terrestre," *Journal de physique, 46,* 428–435.

1799 "Mémoire sur la matière du feu, considéré comme instrument chimique dans les analyses," *Journal de physique, 48,* 345–361.

1799 "Mémoire sur la matière du son," *Journal de physique, 49,* 397–412.

1799 "Sur les genres de la Seiche, du Calmar et du Poulpe vulgairement nommés polypes de mer," *Mémoires de la Société d'Histoire Naturelle de Paris, 1,* 1–25.

1799 "Prodrome d'une nouvelle classification des coquilles, comprenant une rédaction appropriée des caractères génériques, et l'établissement d'un grand nombre de genres nouveaux," *Mémoires de la Société d'Histoire Naturelle de Paris, 1,* 63–91.

1800–1810 *Annuaires météorologiques.* 11 vols. Paris.

1801 *Système des animaux sans vertèbres, ou tableau général des classes, des ordres et des genres de ces animaux; présentant leurs caractères essentiels et leur distribution, d'après la considération de leurs rapports naturels et de leur organisation, et suivant l'arrangement établi dans les galeries du Muséum d'Histoire naturelle, parmi leurs dépouilles conservées; précédé du discours d'ouverture du cours de zoologie, donné dans le Muséum national d'Histoire Naturelle, l'an VIII de la République.* Paris.

1801 "Recherches sur la périodicité présumée des principales variations de l'atmosphère, et sur les moyens de s'assurer de son existence et de sa determination," *Journal de physique, 52,* 296–316.

1802 "Sur les variations de l'état du ciel dans les latitudes moyennes entre l'équateur et le pôle, et sur les principales causes qui y donnent lieu," *Journal de physique, 56,* 114–138.

1802 *Hydrogéologie ou recherches sur l'influence qu'ont les eaux sur la surface du globe terrestre; sur les causes de l'existence du bassin des mers, de son déplacement et de son transport successif sur les différens points de la surface de ce globe; enfin sur les changements que les corps vivans exercent sur la nature et l'état de cette surface.* Paris.

1802 "Considérations sur les observations météorologiques comparatives, et sur leur application," *Annales de statistique, ou journal général d'économie politique, industrielle et commerciale; de géographie, d'histoire naturelle, d'agriculture, de physique, d'hygiène et de littérature, 3,* 58–72.

1802 "Lettre du cit. J.-B. Lamarck, à L.-J. P. Ballois," *Annales de statistique, 3,* 78–82.

1802 "Sur les relations qui existent entre la météorologie-statistique et la météorologie générale, et sur les moyens d'avancer les progrès de l'une et de l'autre," *Annales de statistique, 3,* 300–317.

1802 "Sur les observations et le but de toute météorologie-statistique," *Annales de statistique, 4,* 129–134.

1802 *Recherches sur l'organisation des corps vivans, et particulièrement sur son origine, sur la cause de ses développemens et des progrès de sa composition, et sur celle qui, tendant continuellement à la détruire dans chaque individu, amène nécessairement sa mort; précédé du discours d'ouverture du cours de zoologie, donné dans le Muséum national d'Histoire Naturelle, l'an X de la République.* Paris.

1802–1806 "Mémoires sur les fossiles des environs de Paris, comprenant la détermination des espèces qui apartiennent aux animaux marins sans vertèbres, et dont la plupart sont figurés dans la collection des vélins du Muséum," *Annales du Muséum d'Histoire Naturelle, 1,* 299–312, 383–391, 474–479; *2,* 57–64, 163–169, 217–227, 315–321, 385–391; *3,* 163–170, 266–274, 343–352, 436–441; *4,* 46–55, 105–115, 212–222, 289–298, 429–436; *5,* 28–36, 91–98, 179–188, 237–245, 349–357; *6,* 117–126, 214–221, 337–346, 407–415; *7,* 53–62, 130–140, 231–239, 419–430; *8,* 156–158, 347–356, 461–470.

1803 *Histoire naturelle des végétaux.* vols. 1 and 2 (of 15). Paris.

1803 *Discours d'ouverture d'un cours de zoologie, prononcé en prairial an XI, au Muséum d'Histoire naturelle, sur la question: qu'est-ce que l'espèce parmi les corps vivans?* Paris. Reprinted in A. Giard (ed.), *Discours d'ouverture des cours de zoologie donnés dans le Muséum d'Histoire naturelle (An VIII, An X, An XI et 1806), par J.-B.- Lamarck.* Paris, 1907, pp. 85–105.

1804 "Sur une nouvelle espèce de Trigonie, et sur une nouvelle d'Huître, découvertes dans le voyage du capitaine Baudin," *Annales du Muséum d'Histoire Naturelle, 4,* 351–359.

1805 "Considerations sur quelques faits applicables à la théorie du globe, observés par M. Péron dans son voyage aux terres australes, et sur quelques questions géologiques qui naissent de la connoissance de ces faits," *Annales du Muséum d'Histoire Naturelle, 6,* 26–52.

1806 *Discours d'ouverture du cours des animaux sans vertèbres, prononcé dans le Muséum d'Histoire naturelle en mars 1806.* Paris. In Giard (ed.), *Discours d'ouverture.* Paris, 1907, pp. 107–157.

1809 *Philosophie zoologie, ou exposition des considérations relatives à l'histoire naturelle des animaux; à la diversité de leur organisation et des facultés qu'ils en obtiennent; aux causes physiques qui maintiennent en eux la vie et donnent lieu aux mouvemens qu'ils exécutent; enfin, à celles qui produisent, les unes le sentiment, et les autres l'intelligence de ceux qui en sont doués.* Paris.

1810 "Sur la détermination des espèces parmi les animaux sans vertèbres et particulièrement parmi les mollusques testacés," *Annales du Muséum d'Histoire Naturelle, 15,* 20–26.

1812 *Extrait du cours de zoologie du Muséum d'Histoire naturelle sur les animaux sans vertèbres, présentant la distribution et classification de ces animaux, les caractères des principales divisions et une simple liste des genres, à l'usage de ceux qui suivent ce cours.* Paris.

1815–1822 *Histoire naturelle des animaux sans vertèbres, présentant les caractères généraux et particuliers de ces animaux, leur distribution, leurs classes, leurs familles, leurs genres, et la citation des principales espèces qui s'y rapportent; précédée d'une introduction offrant la détermination des caractères essentiels de l'animal, sa distinction du végétal et des autres corps naturels, enfin, l'exposition des principes fondamentaux de la zoologie.* 7 vols. Paris.

1817 "Conchifères," *Nouveau dictionnaire d'histoire naturelle, 7,* 409–411.

1817 "Conchyliologie," *Nouveau dictionnaire d'histoire naturelle, 7,* 412–428.

1817 "Coquillage," *Nouveau dictionnaire d'histoire naturelle, 7,* 547–556.

1817 "Coquille," *Nouveau dictionnaire d'histoire naturelle, 7,* 556–583.

1817 "Espèce," *Nouveau dictionnaire d'histoire naturelle, 10,* 441–451.

1817 "Faculté," *Nouveau dictionnaire d'histoire naturelle, 11,* 8–18.

1817 "Fonctions organiques," *Nouveau dictionnaire d'histoire naturelle, 11,* 593–596.

1817 "Habitude," *Nouveau dictionnaire d'histoire naturelle, 14,* 128–138.

1817 "Homme," *Nouveau dictionnaire d'histoire naturelle, 15,* 270–276.

1817 "Idée," *Nouveau dictionnaire d'histoire naturelle, 16,* 78–94.

1817 "Imagination," *Nouveau dictionnaire d'histoire naturelle, 16,* 126–132.

1817 "Instinct," *Nouveau dictionnaire d'histoire naturelle, 16,* 331–343.

1817 "Intelligence," *Nouveau dictionnaire d'histoire naturelle, 16,* 344–360.

1817 "Jugement," *Nouveau dictionnaire d'histoire naturelle, 16,* 570–579.

1818 "Météores," *Nouveau dictionnaire d'histoire naturelle, 20,* 416–444.

1818 "Météorologie," *Nouveau dictionnaire d'histoire naturelle, 20,* 451–477.

1818 "Nature," *Nouveau dictionnaire d'histoire naturelle, 22,* 363–399.

1820 *Système analytique des connaissances positives de l'homme restreintes à celles qui proviennent directement ou indirectement de l'observation.* Paris.

Writers other than Lamarck

Adanson, Michel. "Examen de la question, si les espèces changent parmi les plantes; nouvelles expériences tentées à ce sujet," *Mémoires de l'Académie Royale des Sciences* (1769), 383–419.

———. *Familles des plantes.* 2 vols. Paris, 1763.

Arago, François. *Oeuvres de François Arago.* 17 vols. Paris and Leipzig, 1854–1862.

Aron, Jean-Paul. "Les circonstances et le plan de la nature chez Lamarck," *Essais d'epistémologie biologique.* Paris, 1969, pp. 83–98.

Aufrère, Léon. *De Thalès à Davis: le relief et la sculpture de la terre. . . . Soulavie et son secret.* Paris, 1952.

d'Aumont. "Génération," *Encyclopédie, 7* (1757), 559–574.

Bernardin de Saint-Pierre, Jacques-Henri. *Études de la nature.* 4th ed. 5 vols. Paris, 1792.

Bertholon. "Nouvelles expériences sur les effets de l'électricité artificielle & naturelle, appliquée aux végétaux," *Journal de physique, 35* (1789), 401–423.

Blainville, M. H. de. *Histoire des sciences de l'organisation et de leurs progrès, comme base de la philosophie.* Ed. F. L. M. Maupied. 3 vols. Paris, 1847.

Boinet, A. *Catalogue général des manuscrits des bibliothèques publiques de France: Paris, tome II, Muséum d'Histoire Naturelle.* Paris, 1914.

Bonald, L. G. A. de. *Recherches philosophiques sur les premiers objets des connaissances morales.* 2 vols. Paris, 1818.

Bonnet, Charles. *Contemplation de la nature.* New ed. 3 vols. Hamburg, 1782.

———. *Oeuvres d'histoire naturelle et de philosophie.* 18 vols. Neuchatel, 1779–1783.

Bonnet, Ed. "L'herbier de Lamarck: son histoire, ses vicissitudes, son état actuel," *Journal de botanique* (Morot), *16* (1902), 129–138.

Bosc, L. A. G. *Histoire naturelle des coquilles, contenant leur description, les moeurs des animaux qui les habitent et leurs usages.* 5 vols. Paris, an 10 (1802).

Bourdier, Franck. "Lamarck et Geoffroy Saint-Hilaire face au problème de l'évolution biologique," *Revue d'histoire des sciences, 25* (1972), 311–325.

———. "L'homme selon Lamarck," in J. Schiller (ed.), *Colloque international "Lamarck."* Paris, 1971, pp. 137–159.

———. "Quelques aperçus sur la paléontologie évolutive en France avant Darwin," *Bulletin de la Société Géologique de France,* 7th ser. *1* (1959) 882–896.

———. "Trois siècles d'hypothèses sur l'origine et la transformation des êtres vivants (1550–1859)," *Revue d'histoire des sciences, 13* (1960), 1–44.

Bourdier, Franck, and Michael Orliac. "Esquisse d'une chronologie de la vie de Lamarck." Unpublished manuscript, 3rd section, École Pratique des Hautes Études, 22 June 1971.

Bourdon, Isid. *Illustres médecins et naturalistes des temps modernes.* Paris, 1844.

Bowler, Peter J. "Bonnet and Buffon: theories of generation and the problem of species," *Journal of the History of Biology, 6* (1973), 259–281.

Boyer, Ferdinand. "Le Muséum d'Histoire Naturelle à Paris et l'Europe des sciences sous la Convention," *Revue d'histoire des sciences, 26* (1973), 251–257.

Brémond, J., and J. Lassertisseur, "Lamarck et l'entomologie," *Revue d'histoire des sciences, 26* (1973), 231–250.

Buffon. *Les époques de la nature.* Ed. Jacques Roger. Paris, 1962.

———. *Histoire naturelle, générale et particulière.* 44 vols. Paris, 1749–1804.

———. *Oeuvres complètes de Buffon.* Ed. Lamouroux and A. G. Desmarest. 40 vols. Paris, 1824–1831.

———. *Oeuvres complètes de Buffon.* Ed. J.-L. Lanessan. 14 vols. Paris, 1884–1885.

———. *Oeuvres philosophiques de Buffon.* Ed. Jean Piveteau. Paris, 1954.

Burkhardt, Richard W., Jr. "Lamarck, evolution, and the politics of science," *Journal of the History of Biology, 3* (1970), 275–298.

———. "Latreille," *Dictionary of scientific biography, 8* (1973), 48–49.

———. "The inspiration of Lamarck's belief in evolution," *Journal of the History of Biology, 5* (1972), 413–438.

Burlingame, Leslie J. "Lamarck," *Dictionary of scientific biography, 7* (1973), 584–594.

———. "Lamarck's theory of transformism in the context of his views of nature." Ph.D. dissertation, Cornell University, 1973.

Cabanis, P.-J.-G. *Oeuvres philosophiques de Cabanis.* Ed. Claude Lehec and Jean Cazeneuves. 2 vols. Paris, 1956.

———. *Rapports du physique et du moral de l'homme.* 2 vols. Paris, 1802.

Cahn, Théophile. *La vie et l'oeuvre scientifique d'Étienne Geoffroy Saint-Hilaire.* Paris, 1962.

Cannon, H. Graham. "What Lamarck really said," *Proceedings of the Linnean Society of London, 168* (1957), 70–85.

Candolle, Augustin Pyramus de. *Mémoires et souvenirs de Augustin Pyramus de Candolle.* Geneva, 1862.

Catalogue des livres de la bibliothèque du feu M. le Chevalier J.-B. de Lamarck. Paris, 1830.

Clos, Dominique. "Lamarck botaniste, sa contribution à la méthode dite naturelle et à la troisième édition de la *Flore française*," *Mémoires de l'Académie des Sciences, inscriptions et belles-lettres de Toulouse, 8* (1896), 202–225.

Coleman, William. *Georges Cuvier, zoologist: a study in the history of evolution theory.* Cambridge, Mass., 1964.

Crocker, Lester G. "Diderot and eighteenth century French transformism," in Glass et al. (eds.), *Forerunners of Darwin.* Baltimore, 1959, pp. 114–143.

Crosland, Maurice P. *The Society of Arcueil: a view of French science at the time of Napoleon I.* London, 1967.

Cuvier, Georges. "Éloge de M. Lamarck," *Mémoires de l'Académie Royale des Sciences de l'Institut de France, 13* (1835), i-xxxi.

———. "Extrait d'un ouvrage sur les espèces de quadrupèdes dont on a trouvé les ossemens dans l'intérieur de la terre, addressé aux savans et aux amateurs des sciences," *Magasin encyclopédique, 7* (1801), no. 1, 60–82.

———. *Histoire des sciences naturelles, depuis leur origine jusqu-à nos jours, chez tous les peuples connus, professé au Collège de France.* Ed. Magdeleine de Saint-Agy. 5 vols. Paris, 1841–1845.

———. *Leçons d'anatomie comparée.* Ed. C. Duméril. 5 vols. Paris, an 8 (1800)–an 14 (1805).

———. *Leçons d'anatomie comparée.* 2nd. ed. 8 vols. Paris, 1835–1846.

———. "Mémoire sur la structure interne et externe, et sur les affinités des animaux auxquels on a donné le nom de vers," *La décade philosophique, 5* (1795), 385–396.

———. "Mémoire sur les espèces d'éléphans vivantes et fossiles," *Magasin encyclopédique, 2* (1796), no. 3, 440–445; *Mémoires de l'Institut de France (Class. math. phys.), 2* (1799), 1–22.

———. "Mémoire sur l'ibis des anciens Égyptiens," *Annales du Muséum d'Histoire Naturelle, 4* (1804), 116–135.

———. "Nature," *Dictionnaire des sciences naturelles, 34* (1825), 261–268.

———. "Nouveaux système des animaux sans vertèbres," *Magasin encyclopédique, 6* (1801), no. 6, 387–388.

———. "Rapport de l'Institut national . . . sur un ouvrage de M. André, ayant pour titre: théorie de la surface actuelle de la terre," *Journal des mines, 21* (1807), 413–430.

———. *Recherches sur les ossemens fossiles de quadrupèdes, où l'on rétablit les caractères de plusieurs espèces d'animaux que les révolutions du globe paroissent avoir détruites.* 4 vols. Paris, 1812.

———. *Recueil des éloges historiques lus dans les séances publiques de l'Institut de France.* New ed. 3 vols. Paris, 1861.

———. *Le règne animal distribué d'après son organisation, pour servir de base à l'histoire naturelle des animaux et d'introduction à l'anatomie comparée.* 4 vols. Paris, 1817.

———. *Tableau élémentaire d'histoire naturelle.* Paris, an 6 (1798).

Dagognet, F. *Le catalogue de la vie.* Paris, 1970.

Darnton, Robert. *Mesmerism and the end of the enlightenment in France.* Cambridge, Mass., 1968.

Darwin, Charles. *On the origin of species.* Facsimile of the first edition. Cambridge, Mass., 1966.

———. *More letters of Charles Darwin.* Ed. Francis Darwin. 2 vols. London, 1903.

Darwin, Erasmus. *Les amours des plantes.* Tr. J. P. F. Deleuze. Paris, an 8 (1800).

———. *Zoonomia; or, the laws of organic life.* 2 vols. London, 1794–1796.

Daubenton, L.-J.-M. *Encyclopédie méthodique: histoire naturelle des animaux.* Paris, 1782.

———. "Histoire naturelle," *Encyclopédie ou Dictionnaire raisonné des sciences, des arts, et des métiers . . . , 8* (1755), 225–230.

———. "Programme; sur la définition et les limites de l'histoire naturelle," in *Séances des écoles normales, recueillies par des sténographes et revues par les professeurs.* New ed. 8 vols. Paris, 1800.

———. "Sur la division méthodique des animaux," *Séances des écoles normales, 6,* 104–135.

———. "Sur la nomenclature méthodique de l'histoire naturelle," *Séances des écoles normales, 1,* 425–444.

———. "Sur la physiologie des végétaux, comparée à celle des animaux," *Séances des école normales, 5,* 269–278.

———. "Sur les couches du globe de la terre," *Séances des écoles normales, 2,* 265–290.

———. "Sur les rapports que l'on a recherchés entre les corps bruts et les corps organisés," *Séances des écoles normales, 4,* 3–13.

———. "Sur les voyages et les théories des naturalistes," *Séances des écoles normales, 2,* 171–188.

Daudin, Henri. *Cuvier et Lamarck: Les classes zoologiques et l'idée de série animale (1790–1830).* 2 vols. Paris, 1926.

———. *De Linné à Jussieu: méthodes de classification et idée de série en botanique et en zoologie (1740–1790).* Paris, n.d. (1926).

Dehérain, Henri. *Catalogue des manuscrits du fonds Cuvier (travaux et correspondance scientifiques) conservés à la bibliothèque de l'Institut de France.* Paris, 1908.

Delamétherie, Jean-Claude. *Considérations sur les êtres organisés.* 3 vols. Paris, 1804–1806.

———. *Leçons de géologie données au Collège de France.* 3 vols. Paris, 1816.

———. *De la nature des êtres existans, ou principes de la philosophie naturelle.* Paris, 1805.

———. *Théorie de la terre.* 3 vols. Paris, 1795.

———. *Vues physiologiques sur l'organisation animale et végétale.* Amsterdam, 1780.

Deleuze, Joseph Philippe François. *Histoire et description du Muséum Royal d'Histoire Naturelle.* 2 vols. Paris, 1823.

Denys de Montfort, Pierre. *Conchyliologie systématique, et classification méthodique des coquilles.* 2 vols. Paris, 1808–1810.

———. *Histoire naturelle, générale et particulière des mollusques, animaux sans vertèbres et à sang blanc.* 5 vols. Paris, an 10 (1802)–an 13 (1805).

Derham, William. *Physico-theology, or a demonstration of the being and attributes of God, from his work of creation.* 3rd ed. London, 1714.

Desfontaines, René-Louiche. "Histoire naturelle. Cours de botanique élémentaire et de physique végétale. Discours d'ouverture," *La décade philosophique, 5* (1795), 450–461, 513–520.

———. *Tableau de l'école de botanique du Jardin du Roi.* 2nd ed. Paris, 1815.

Deshayes, G. P. *Description des animaux sans vertèbres découverts dans le bassin de Paris.* Paris, 1860.

Desmarest, A. and Lesueur. "Note sur le Botrylle étoilé (*Botryllus stellatus*) Pall.," *Bulletin des sciences* (1815), 74–78.

Desmarest, Nicolas. *Encyclopédie méthodique: géographie-physique.* 5 vols. Paris, 1794–1828.

Diderot, Denis. *Oeuvres philosophiques de Diderot.* Ed. Paul Vernière. Paris, 1964.

Duchesne, Antoine-Nicolas. "Classification des habitans des eaux," *Magasin encyclopédique, 2* (1796), no. 3, 300–301.

———. *Histoire naturelle des fraisiers.* Paris, 1766.

———. "Sur le fraisier de Versailles," *Journal d'histoire naturelle, 2* (1792), 343–347.

———. "Sur les rapports entres les êtres naturels," *Magasin encyclopédique, 1* (1795), no. 6, 289–294.

Dunmore, John. *French explorers in the Pacific.* 2 vols. Oxford, 1965–1969.

Dupuy, Paul. *L'école normale de l'an III.* Paris, 1895.

Egerton, Frank N. "Changing concepts of the balance of nature," *Quarterly Review of Biology, 48* (1973), 322–350.

Farber, Paul. "Buffon and the concept of species," *Journal of the History of Biology, 5* (1972), 259–284.

Farley, John. "The spontaneous generation controversy (1700–1860): the origin of parasitic worms," *Journal of the History of Biology, 5* (1972), 95–125.

Faujas de Saint-Fond, Barthélemy. *Essai de géologie, ou mémoires pour servir à l'histoire naturelle du globe.* 2 vols. Paris, 1803–1809.

———. *Histoire naturelle de la montagne de Saint-Pierre de Maestricht.* Paris, an 7 (1799).

Fischer, Gotthelf, *Das Nationalmuseum der Naturgeschichte zu Paris.* 2 vols. Frankfurt, 1802.

Fortis, Albert. *Mémoires pour servir à l'histoire naturelle et principalement à l'oryctographie de l'Italie, et des pays adjacens.* 2 vols. Paris, 1802.

Foucault, Michel. *The order of things: an archeology of the human sciences.* New York, 1971.

Franklin, le Roy, De Bory, Lavoisier, and Bailly. "Exposé des expériences qui ont été faites pour l'examen du magnétisme animal," *Histoire de l'Académie Royale des Sciences* (1784), 6–15.

Gall, F. J., and J. C. Spurzheim. *Des dispositions innés de l'âme et de l'esprit, du matérialisme, du fatalisme et de la liberté morale, avec des réflexions sur l'éducation et sur la législation criminelle.* Paris, 1811.

Geoffroy Saint-Hilaire, Étienne. *Fragments biographiques, précédés d'études sur la vie, les ouvrages et les doctrines de Buffon.* Paris, 1838.

———. "Recherches sur l'organisation des gavials," *Mémoires du Muséum d'Histoire Naturelle, 12* (1825), 97–155.

———. "Rapport fait à l'Académie des Sciences sur un mémoire de M. Roulin, ayant pour titre: sur quelques changemens observés dans les animaux domestiques transportés de l'ancien monde dans le nouveau continent," *Mémoires du Muséum d'Histoire Naturelle, 17* (1828), 201–208.

———. "Sur le degré de l'influence du monde ambiant pour modifier les formes animales; question intéressant l'origine des espèces téléosauriennes et successivement celle des animaux de l'époque actuelle," *Mémoires de l'Académie Royale des Sciences,* 2nd ser. *12* (1833), 63–92.

Geoffroy Saint-Hilaire, Étienne, and Georges Cuvier. "Mémoire sur les orang-outangs," *Magasin encyclopédique 1* (1795), no. 3, 451–463; *Journal de physique, 46* (1798), 185–191.

Geoffroy Saint-Hilaire, Isidore. *Vie, travaux et doctrine scientifique d'Étienne Geoffroy Saint-Hilaire.* Paris, 1847.

"Georges Cuvier: Journées d'études organisées par l'Institut d'histoire des sciences de l'Université de Paris," *Revue d'histoire des sciences, 23* (1970), 7–92.

Gillispie, Charles C. "Lamarck and Darwin in the history of science," in Glass et al (eds.), *Forerunners of Darwin,* Baltimore, 1959, pp. 265–291.

———. *The edge of objectivity: an essay in the history of scientific ideas.* Princeton, 1960.

———. "The *Encyclopédie* and the Jacobin philosophy of science; a study in ideas and consequences," in Marshall Claggett (ed.), *Critical problems in the history of science.* Madison, 1959, pp. 255–289.

———. "The formation of Lamarck's evolutionary theory," *Archives internationales d'histoire des sciences, 9* (1956), 323–338.

Glass, Bentley. "Heredity and variation in the eighteenth century concept of species," in Glass et al. (eds.), *Forerunners of Darwin,* Baltimore, 1959, pp. 144–172.

———. "Maupertuis, pioneer of genetics and evolution," in Glass et al. (eds.), *Forerunners of Darwin.* Baltimore, 1959, pp. 51–83.

Glass, Bentley, Owsei Temkin, and William L. Strauss, Jr. (eds.). *Forerunners of Darwin.* Baltimore, 1959.

Gohau, Gabriel. "Le cadre minéral de l'évolution Lamarckienne," in J. Schiller (ed.), *Colloque international "Lamarck."* Paris, 1971, pp. 105–133.

Gould, Stephen Jay. "Trigonia and the origin of species," *Journal of the History of Biology, 1* (1968), 41–56.

Greene, John C. *The death of Adam*. Ames, Iowa, 1959.

———. "The Kuhnian paradigm and the Darwinian revolution in natural history," in Duane H. D. Roller (ed.), *Perspectives in the history of science and technology*. Norman, Oklahoma, 1971, pp. 3–25.

Guyénot, Émile. *Les sciences de la vie aux xvii*e *et xviii*e *siècles: l'idée d'évolution*. Paris, 1941.

Haber, Francis C. *The age of the world: Moses to Darwin*. Baltimore, 1959.

———. "Fossils and the idea of a process of time in natural history," in Glass et al. (eds.), *Forerunners of Darwin*, Baltimore, 1959, pp. 222–261.

Hahn, Roger. *The anatomy of a scientific institution: the Paris Academy* of Sciences, *1666–1803*. Berkeley, 1971.

———. "L'autobiographie de Lacépède retrouvée," *Dix-huitième siècle, 7* (1975), 49–85.

———. "Sur les débuts de la carrière scientifique de Lacépède," *Revue d'histoire des sciences, 27* (1974), 347–353.

Hall, Thomas S. *Ideas of life and matter*. 2 vols. Chicago, 1969.

Haller, Baron Albertus. *First lines of physiology*. Reprint of the 1786 edition. New York and London, 1966.

Hamy, Ernest-Théodore. *Les débuts de Lamarck, suivis de recherches sur Adanson, Jussieu, Pallas, Geoffroy Saint-Hilaire, Georges Cuvier, etc.* Paris, n.d. (1909).

———. "Les derniers jours du Jardin du Roi et la fondation du Muséum d'Histoire Naturelle," *Centenaire de la fondation du Muséum d'Histoire Naturelle*. Paris, 1893, pp. 1–162.

Hodge, M. J. S. "La métaphysique de Lamarck d'après un opuscule retrouvé," *Revue d'histoire des sciences, 26* (1973), 223–229.

———. "Lamarck's science of living bodies," *British Journal of the History of Science, 5* (1971), 323–352.

———. "Species in Lamarck," in J. Schiller (ed.), *Colloque international "Lamarck."* Paris, 1971, pp. 31–46.

Hofsten, Nils von. "Linnaeus's conception of nature," *Kungl. Vetenskaps-Societetens Årsbok* (1957), 65–105.

Holbach. "Fossiles," *Encyclopédie, 7* (1757), 209–211.

———. "Préface du traducteur," in Jean-Gotlob Lehmann, *Traités de physique, d'histoire naturelle, de minéralogie et de métallurgie*. 3 vols. Paris, 1759.

Hooykaas, Reijer. *Natural law and divine miracle: a historical-critical study of the principle of uniformity in geology, biology, and theology*. Leiden, 1959.

Ingen-Housz. "Lettre . . . au sujet de l'influence de l'électricité atmosphérique sur les végétaux," *Journal de physique, 32* (1788), 321–337.

Jussieu, Antoine-Laurent de. "Examen de la famille des Renoncules," *Mémoires de l'Académie Royale des Sciences* (1773), 214–240.

———. "Exposition d'un nouvel ordre de plantes adopté dans les démonstrations du Jardin Royal," *Mémoires de l'Académie Royale des Sciences* (1774), 175–197.

———. "Notice sur l'expédition à la Nouvelle-Hollande entreprise pour des recherches de géographie et d'histoire naturelle," *Annales du Muséum d'Histoire Naturelle, 5* (1804), 1–11.

King, Lester. "Stahl and Hoffmann: A study in eighteenth century animism," *Journal of the History of Medicine, 19* (1964), 118–130.

———. *The growth of medical thought.* Chicago, 1963.

Kirby, William. *On the power, wisdom and goodness of God, as manifested in the creation of animals, and in their history, habits and instincts.* Philadelphia, 1837.

Kirby, William, and William Spence. *An introduction to entomology, or elements of the natural history of insects.* 4 vols. London, 1818–1826.

Lacépède, Bernard-Germain-Étienne de la Ville-sur-Illon. "Considérations sur les parties du globe dans lesquelles on n'a pas encore pénétré," *Séances des écoles normales, 8,* 195–207.

———. *Discours d'ouverture et de clôture du cours d'histoire naturelle, donné au Muséum en l'an VI.* Paris, 1798.

———. *Discours d'ouverture et de clôture du cours d'histoire naturelle, donné dans le Muséum National d'Histoire Naturelle, l'an VII de la République.* Paris, an 7 (1799).

———. *Histoire naturelle des poissons.* 5 vols. Paris, 1798–1802.

———. *Les âges de la nature et histoire de l'espèce humaine.* Paris, 1830.

———. *Oeuvres du comte de Lacépède.* New ed. 11 vols. Paris, 1826–1833.

———. *Physique générale et particulière.* 2 vols. Paris, 1782.

Lacépède, Cuvier, and Lamarck. "Rapport des professeurs du Muséum sur les collections d'histoire naturelle rapportées d'Égypte," *Annales du Muséum d'Histoire Naturelle, 1* (1802), 234–241.

Laissus, Yves. "Les cabinets d'histoire naturelle," in René Taton (ed.), *Enseignement et diffusion des sciences en France au xviii*[e] *siècle.* Paris, 1964, pp. 659–712.

Lamouroux. *Histoire des polypes coralligènes flexibles.* Caen, 1861.

Lamy, Ed. *Les cabinets d'histoire naturelle en France au xviii*[e] *siècle et le cabinet du roi (1635–1793).* Paris, n.d. (1930).

———. "Les conchyliologistes Bruguière et Hwass," *Journal de conchyliologie, 74* (1930), 42–43.

Landrieu, Marcel. *Lamarck, le fondateur du transformisme.* Paris, 1909.

Larson, James L. *Reason and experience: the representation of natural order in the work of Carl von Linné.* Berkeley, 1971.

———. "The species concept of Linnaeus," *Isis, 59* (1968), 291–299.

Lassus. "Sciences physiques," *Magasin encyclopédique, 5* (1799), no. 1, 233–234.

Latreille, Pierre-André. *Considérations générales sur l'ordre naturel des animaux composant les classes des crustacés, des arachnides et des insectes.* Paris, 1810.

———. *Cours d'entomologie, ou de l'histoire naturelle des crustacés, des arachnides, des myriapodes et des insectes.* Paris, 1831.

Lavoisier, A. L. "Observations générales, sur les couches modernes horizontales, qui ont été déposées par la mer, sur les conséquences qu'on peut tirer de leurs dispositions, relativement à l'ancienneté du globe terrestre," *Mémoires de l'Académie Royale des Sciences* (1789 [1793]), 351–371.

———. *Traité élémentaire de chimie.* 2 vols. Paris, 1789.

Lesueur, Ch. Alex. "Mémoire sur l'organisation des Pyrosomes," *Bulletin des sciences* (1815), 70–74.

Limoges, Camille. *La sélection naturelle.* Paris, 1970.

Linné, Charles. *Philosophie botanique.* Tr. F. A. Quesné. Paris, 1788.

———. *Revue générale des écrits de Linné.* Ed. Richard Pulteney. Tr. L. A. Millin de Grandmaison. 2 vols. London and Paris, 1789.

———. *Système de la nature.* 13th ed. Ed. J. F. Gmelin. Tr. Vanderstegen de Putte. Bruxelles, 1793.

Lovejoy, Arthur O. "Buffon and the problem of species," in Glass et al. (eds.), *Forerunners of Darwin.* Baltimore, 1959, pp. 84–113.

———. *The great chain of being.* Cambridge, Mass., 1936.

Lyell, Charles. *Principles of geology, being an attempt to explain the former changes of the earth's surface by reference to causes now in operation.* 3 vols. London, 1830–1833.

Maillet, Benoît de. *Telliamed, ou entretiens d'un philosophe indien avec un missionaire francois sur la diminution de la mer, la formation de la terre, l'origine de l'homme, &c.* 2 vols. Amsterdam, 1748.

Marchant, James. "Observations sur la nature des plantes," *Mémoires de l'Académie Royale des Sciences* (1719), 59–66.

Maupertuis, Pierre Louis Moreau de. *Oeuvres.* 4 vols. Lyon, 1768. Reprinted Hildesheim, 1965–1974.

Mayr, Ernst. "Illiger and the biological species concept," *Journal of the History of Biology, 1* (1968), 163–178.

———. "Lamarck revisited," *Journal of the History of Biology, 5* (1972), 55–94.

———. "Species concepts and definitions," in Mayr (ed.), *The species problem.* Washington, 1957.

Mornet, Daniel. *Les sciences de la nature en France au xviii*e *siècle.* Paris, 1911.

Musée d'Histoire Naturelle de Génève. *Catalogue illustré de la collection Lamarck.* Geneva, 1910–1918.

Nicard, Pol. *Étude sur la vie et les travaux de M. Ducrotay de Blainville.* Paris, 1890.

Nisbet, Robert A. *Social change and history: aspects of the Western theory of development.* London, Oxford, and New York, 1969.

Nollet, l'Abbé. *Recherches sur les causes particulières des phénomènes électriques, et sur les effets nuisibles ou avantageux qu'on peut en attendre.* Paris, 1749.

"Objets d'histoire naturelle recueillis en Hollande," *Décade philosophique, 5* (an 3 [1795]), 535–536.

Olby, Robert. *Origins of Mendelism.* New York, 1966.

Olivier, G. A. *Encyclopédie méthodique: histoire naturelle des insectes.* Vol. 7. Paris, 1792.

———. "Mémoire sur l'utilité de l'étude des insectes, relativement à l'agriculture et aux arts," *Journal d'histoire naturelle, 1* (1792), 33–56.

———. *Voyage dans l'Empire othoman, l'Égypte et la Perse.* 6 vols. Paris, 1801–1807.

Osborn, H. F. *From the Greeks to Darwin.* 2nd ed. New York, 1929.

Ostoya, Paul. *Les théories de l'évolution.* Paris, 1951.

Perrier, Edmond. *La philosophie zoologique avant Darwin.* Paris, 1896.

Picavet, F. *Les idéologues: essai sur l'histoire des idées et des théories scientifiques, philosophiques, religieuses, etc. en France depuis 1789.* Paris, 1891.

Pilet, P. E. "Charles Bonnet," *Dictionary of scientific biography, 2* (1970), 286–287.

Pluche, l'Abbé. *La spectacle de la nature, ou entretiens sur les particularités de l'histoire naturelle qui ont paru les plus propres à rendre les jeunes gens curieux & à leur former l'esprit.* 8 vols. Paris, 1782–1787.

Poliakov, I. M. *J.-B. Lamarck and the theory of the evolution of the organic world.* In Russian. Moscow, 1962.

Priestley, Joseph. *The history and present state of electricity, with original experiments.* 3rd ed. 2 vols. London, 1755. Reprinted New York, 1966.

Quatrefages, A. de. *Charles Darwin et ses précurseurs français: étude sur le transformisme.* Paris, 1870.

Quatremère Disjonval. *De l'aranéologie, ou sur la découverte du rapport constant entre l'apparition ou la disparition, le travail ou le moins d'étendue des toiles et des fils d'attaches des araignées des differentes espèces.* Paris, an 5 (1797).

Rappaport, Rhoda. "G.-F. Rouelle: an eighteenth century chemist and teacher," *Chymia, 6* (1960), 68–101.

Raspail, F.-V. "Nécrologie; parallèle," *Annales des sciences d'observation, 3* (1830), 159–160.

Reynier, Louis. "De l'influence du CLIMAT sur la forme et la nature des végétaux," *Journal d'histoire naturelle, 2* (1792), 101–148.

———. "Mémoire pour servir à l'histoire de la marchant variable," *Journal de physique, 30* (1787), 171–174.

———. "Relatif à la formation des corps, par la simple agrégation de la matière organisée," *Journal de physique, 31* (1787), 102–108.

———. "Sur la cristallisation des êtres organisés," *Journal de physique, 33* (1788), 215–217.

———. "Sur la nature du feu," *Journal de physique, 36* (1790), 94–98.

Ritterbush, Philip C. *Overtures to biology: the speculations of eighteenth-century naturalists.* New Haven, 1964.

Robinet, Jean-Baptiste-René. *De la nature.* 4 vols. Amsterdam, 1761–1766.

Rodig. *Lebende Natur.* Leipzig, 1801.

Roger, Jacques. *Les sciences de la vie dans la pensée française du xviii*[e] *siècle: la génération des animaux de Descartes à l'Encyclopédie.* Paris, 1963.

Roule, Louis. *Lamarck et l'interprétation de la nature.* Paris, 1927.

Rousseau, J. J. *Lettres élémentaires sur la botanique,* in *Oeuvres complètes de J. J. Rousseau,* vol. 5. Paris, 1874.

Russell, E. S. *Form and function.* London, 1916.

Savigny, M.-J.-C. *Mémoires sur les animaux sans vertèbres.* 2nd part. Paris, 1816.

Schiller, J. "L'échelle des êtres et la série chez Lamarck," in Schiller (ed.), *Colloque international "Lamarck."* Paris, 1971, pp. 87–103.

———. "Physiologie et classification dans l'oeuvre de Lamarck," *Histoire et biologie, 2* (1969), 35–57.

———. (ed.) *Colloque international "Lamarck."* Paris, 1971.

Schofield, Robert E. *Mechanism and materialism: British natural philosophy in an age of reason.* Princeton, 1970.

Séances des écoles normales, recueillies par des sténographes et revues par les professeurs. New ed. 8 vols. Paris, 1800.

Sherborn, C. Davies, and B. B. Woodward. "On the dates of publication of the natural history portions of the 'Encyclopédie méthodique,'" *Annals and Magazine of Natural History, 17* (1906), 577–582.

Sigaud de la Fond. *Précis historique et expérimental des phénomènes électriques.* 2nd ed. Paris, 1785.

Simpson, George Gaylord. *This view of life.* New York, 1964.

Smith, James Edward. *A sketch of a tour on the continent, in the years 1786 and 1787.* 3 vols. London, 1793.

Société d'Histoire Naturelle de Paris. *Actes, 1* (1792).

———. *Mémoires, 1* (1799).

Soulavie, J. L. Giraud. *Histoire naturelle de la France méridionale.* 8 vols. Paris (vols. 1, 3–8), 1780–1784. Nismes (vol. 2), 1780.

Spallanzani. *Opuscules de physique, animale et végétale.* Tr. Jean Senebier. 3 vols. Pavia, 1787.

Stafleu, Frans. "Lamarck: the birth of biology," *Taxon, 20* (1971), 397–442.

———. "L'Héritier de Brutelle, the man and his work," in Charles-Louis L'Héritier de Brutelle, *Sertum Anglicum, 1788.* Facsimile edition. Pittsburgh, 1963, pp. xiii–lxvi.

Stauffer, Robert Clinton. "Ecology in the long manuscript version of Darwin's *Origin of Species* and Linnaeus' *Oeconomy of Nature,*" *Proceedings of the American Philosophical Society, 104* (1960), 325–341.

Tessier. "Mémoire sur l'importation et les progrès des arbres à épicerie dans les colonies françoises," *Mémoires de l'Académie Royale des Sciences* (1789), 585–596.

Toaldo. "Essai de météorologie appliquée à l'agriculture," *Journal de physique, 10* (1777), 249–279, 333–367.

Topsent, E. "Éponges de Lamarck conservées au Muséum de Paris," *Archives du Muséum d'Histoire Naturelle,* 6th. ser. *5* (1930), 1–56; *8* (1932), 61–124; *10* (1933), 1–60.

Tourlet. Review of Lamarck, *Recherches sur l'organisation des corps vivans,* in *Gazette nationale ou le moniteur universel,* 15 September 1802 (28 fructidor an 10), 1462.

———. Review of Lamarck, *Philosophie zoologique,* in *Gazette nationale ou le moniteur universel,* 18 October 1809, 1154–1156; 24 October 1809, 1178–1180.

Tschoudi, Baron de. *De la transplantation, de la naturalisation et du perfectionnement des végétaux.* London and Paris, 1778.

Vicq-d'Azyr. *Oeuvres de Vicq-d'Azyr.* Ed. Jacq. L. Moreau. 6 vols. Paris, 1805.

Viénot, John. *Le Napoléon de l'intelligence: Georges Cuvier, 1769–1832.* Paris, 1932.

Weiner, Dora B. *Raspail, scientist and reformer.* New York and London, 1968.

Wilkie, J. S. "Buffon, Lamarck, and Darwin: the originality of Darwin's theory of evolution," in P. R. Bell (ed.), *Darwin's biological work: some aspects reconsidered.* Cambridge, 1959, pp. 262–307.

———. "The idea of evolution in the writings of Buffon," *Annals of Science, 12* (1956), 48–62, 212–227, 255–307.

Zirkle, Conway. "The early history of the idea of acquired characters and of pangenesis," *Transactions of the American Philosophical Society,* n.s. *35* (1946), 91–151.

Index